D1275673

ENVIRONMENTAL ACTIVISM

A Reference Handbook

Books in the Contemporary World Issues series address vital issues in today's society such as genetic engineering, pollution, and biodiversity. Written by professional writers, scholars, and nonacademic experts, these books are authoritative, clearly written, up-to-date, and objective. They provide a good starting point for research by high school and college students, scholars, and general readers as well as by legislators, businesspeople, activists, and others.

Each book, carefully organized and easy to use, contains an overview of the subject, a detailed chronology, biographical sketches, facts and data and/or documents and other primary-source material, a directory of organizations and agencies, annotated lists of print and nonprint resources, and an index.

Readers of books in the Contemporary World Issues series will find the information they need in order to have a better understanding of the social, political, environmental, and economic issues facing the world today.

ENVIRONMENTAL ACTIVISM

A Reference Handbook

Jacqueline Vaughn Switzer

ABC-CLIO

Santa Barbara, California
Denver, Colorado
Oxford, England

Copyright © 2003 by Jacqueline Vaughn Switzer

Library of Congress Cataloging-in-Publication Data
Switzer, Jacqueline Vaughn.
  Environmental activism : a reference handbook / Jacqueline Vaughn Switzer.
    p. cm. — (Contemporary world issues)
Includes bibliographical references and index.
  ISBN 1-57607-901-5 (acid-free paper)
    1. Environmentalism—Handbooks, manuals, etc.   I. Title.   II. Series.

GE195 .S88 2003
363.7'0525—dc21

                                                          2002153377

07 06 05 04 03   10 9 8 7 6 5 4 3 2 1

This book is also available on the World Wide Web as an e-book. Visit abc-clio.com for details.

ABC-CLIO, Inc.
130 Cremona Drive, P.O. Box 1911
Santa Barbara, California 93116–1911

This book is printed on acid-free paper ∞.
Manufactured in the United States of America

For the newest generation of environmentalists:
Nathan Stuart Crandall
Joe Martin Nie

Contents

Preface

Over the past four decades, concerns about the environment and its protection have been a consistent element of public discussion and opinion, both in the United States and around the globe. Often, sentiments about what governments ought to do to preserve natural resources and the elements of the global commons such as the oceans and atmosphere are voiced by interest groups and social movements. Sometimes those voices express themselves through the lobbying of members of Congress, state legislators, or members of a local city council. The voices may be shrill and loud, heard through protests and demonstrations of support or opposition. Whatever the tactic and strategy, environmental activists are a fixture in the political process, with varying degrees of success and power.

This book serves as a reference for individuals who want to gain a better understanding of how environmental activism has developed from a historical perspective, placing that activism in the context of contemporary issues, organizations, and conflicts. It is designed to serve as a starting point for further research rather than providing in-depth coverage of specific statutes or programs. It does provide an overview of myriad groups and their strategies, both in the United States and abroad. Whenever possible, the book chronicles both the major events that sparked environmental activism and the nature of that activism, since the two are inextricably linked. For example, the 1969 oil spill off the coast of Santa Barbara, California, was the reason for the formation of Get Oil Out! (GOO!), one of the first grassroots organizations to address marine pollution. GOO! would not have existed without the oil spill and the accompanying media coverage that made a local issue a global one.

The book also seeks to align individuals with activist organizations. Denis Hayes, for example, was a student when he joined forces with Wisconsin senator Gaylord Nelson in organizing the observance of the first Earth Day in 1970. What began as a modest way of publicizing environmental problems has now turned into a major organization that helps coordinate Earth Day events around the world.

Key to understanding this subject is the realization that toward the end of the twentieth century, environmental attitudes and perspectives spread to become global, rather than focused on local or national concerns. Worldwide groups such as Greenpeace and the Rainforest Action Network provide support and resources for advocacy organizations around the world.

Chapter 1 begins with an overview of environmental activism, starting with the formation of the early organizations that were founded around 1900 and leading up to current controversies, including what has come to be known as environmental opposition. It explains how there are distinct and identifiable periods of engagement that differentiate one period of activism from another. The chapter goes on to cover the various types of groups that are active today and the strategies and tactics they use to promote their cause and gain public attention and support.

To illustrate the various ways activists become involved, Chapter 2 focuses on specific controversies, first exploring the issue, then going on to an explanation of the groups involved and their positions. Some problems, such as the drought in Oregon, involve government agencies in conflict with small, rural groups with limited resources and expertise. The chapter goes on to explore the topic of ecoterrorism and the role of violent protest and "ecotage" as a strategy. In Warren County, North Carolina, residents in one of the poorest counties in the state formed a coalition of white landowners and black civil rights leaders to oppose the dumping of chemical wastes in their area. And an unusual coalition of rock climbers, off-highway vehicle users, photographers, tourists, and hikers has joined together to oppose the imposition of government fees to enter public lands.

As evidence of the spread of environmental activism around the globe, Chapter 3 looks at each of the world's continents to survey where advocates for conservation have been successful and where their efforts are still in the beginning stages of development. In some cases, such as Latin America, activists are often silenced through violence, whereas groups in Western Europe are now well entrenched in the political process.

Chapter 4 provides a detailed chronology of activist events from 1900 through 2002, identifying benchmark activities such as the formation of interest groups, noteworthy legislative victories, and landmark court decisions. Chapter 5 follows with profiles of twenty-five individuals and organizations who are exemplary of the differing styles and personalities of some of the world's foremost activists. Some, like Velma Johnston, also known as Wild Horse Annie, are not well known despite their accomplishments. Others, like Chico Mendes, are famous for their role in shaping environmental policy in Latin America. All, however, leave a legacy of commitment and dedication.

One of the best ways to understand group activities is by reading the reports, documents, cases, and testimony of environmental activists and the organizations they represent. Chapter 6 uses primary source material to explain such group strategies as litigation and provides insight into the activists' philosophy and missions. The voices of the activists themselves provide one of the most accurate explanations of what they believe and why they participate in group activities.

The final two chapters are designed to serve as a resource base for those seeking more information about the groups and conflicts described in the first six chapters. There is an extensive directory of environmental activist organizations in Chapter 7, followed in Chapter 8 by print and nonprint sources that are easily accessed through public and university libraries or over the Internet. Chapter 8 includes books, journals, electronic databases, reference books and directories, and a lengthy list of videotapes and their distributors.

The preparation of this book has been an exciting adventure, completed with the assistance of my graduate research assistant at Northern Arizona University, Emily Lethenstrom. I hope that she enjoyed doing the research and drafting portions of the text as much as you will enjoy reading this book and that it will serve as a valuable reference point for further research, debate, and writing.

—*Jacqueline Vaughn Switzer*

1

Environmental Activism in Context

Contemporary public opinion polls show that citizens are over-whelmingly in support of environmental protection. Although other issues such as crime, education, health care, and the economy may rate higher on the political agenda, it is clear that we want and expect our leaders to take the necessary steps to guarantee clean air and water, reduce toxic hazards, restore and protect endangered species and their habitats, manage waste, and ensure environmental health.

But governments depend on individuals to help shape environmental policies and look toward organized groups and their leaders to determine how to prioritize environmental problems and solutions. Under the concept of pluralism, the voices of the people are expressed through groups that represent their views and serve as a conduit to elected officials.

The mechanism for doing that most often takes the form of an interest group, which can comprise a small core of local grass-roots activists or an international organization with a global agenda and thousands of members. Each group, large or small, develops an identity, a mission statement or plan of action, specific strategies, and resources. In order to understand environmental activism today, it is useful to track the development of groups in the United States over the past century. Although some organizations, such as the Sierra Club, had a presence prior to 1900, it is mostly within the past five decades that groups have become part of a distinguishable social movement.

This chapter begins by chronicling the historical context of environmental activism since 1900, exploring how there have been distinct and identifiable periods of engagement with the government and the public. This discussion is followed by an explanation of the various types of activist organizations, their underlying philosophies, and the strategies that are commonly used by groups. Although environmental issues are global, this book is focused primarily on those activist groups that have established a presence in the United States. There is, however, coverage of others whose interests are almost totally limited to a local or regional problem. The result is an overview of how environmental activism has been expressed over the past 100 years.

1900–1945

It has been argued that the beginning of the conservation movement was actually a part of a general backlash in the United States toward the excesses and wastes of the industrial age. The social problems of the late nineteenth century, and the Progressive Era that followed it, led to a call for a return to the simple values of the past. At the same time, Americans realized that they no longer could look to the West as their refuge and that the country's natural resources would have to be utilized responsibly (Kline, 2000).

The best example of that change in view is the debate between preservationists and conservationists, respectively exemplified by John Muir and Gifford Pinchot. Muir had come to California as a young man and had a deep belief that wilderness should exist for its own sake. He sought a political system that would preserve wilderness areas in their natural, pristine state to avoid having them exploited for their resources. In 1892, he joined with Robert Underwood Johnson to found the Sierra Club and, in 1903, lobbied President Theodore Roosevelt to protect Yosemite Valley and its surrounding ecosystem. This decision went beyond simply designating a national park, since it recognized the need to preserve both federal and private lands (Cohen, 1988; Dowie, 1995).

Pinchot served as the leader of the conservation movement, which called for the wise use of the nation's resources. As a French-trained forester, Pinchot believed that the forests were to be used, not preserved as Muir had argued. Pinchot was also a close confidant of President Roosevelt and convinced the president that it was the responsibility of the federal government, rather than private landowners, to regulate natural resources. This philosophy, also called Pinchotism, was heavily criticized by westerners who opposed any form of federal control and intervention (Nash, 1968).

Although there had been a wave of enthusiasm for preservation during the Progressive Era, the designation of forest preserves by Presidents Benjamin Harrison and Grover Cleveland led to protests throughout the West. In June 1907, at the Public Lands Convention in Denver, Colorado, nearly 1,000 delegates protested the federal government's control over public lands and sought to have the lands turned over to the states. The protests led to the formation of the National Public Domain League and the Western Conservation League (1909) to attack Pinchot and seek greater state control of public lands. The protests ended with the advent of World War I, as the country turned its attention elsewhere (Hays, 1959).

During this same period, there developed an early awareness of pollution—an awareness that would serve as the backdrop for initial attempts at protective environmental legislation. For example, the Smoke Prevention Association of America (established in 1907) grew out of the late-nineteenth-century sanitary movement. The health and property damage caused by air pollution was among the most visible environmental problems of this era, and one that could not be ignored or avoided. A similar problem—water quality—was addressed by groups such as the National Coast Anti-Pollution League, formed in 1922 to monitor and prevent water pollution. Women, "the nation's housekeepers," played an important role in calling for urban cleanup efforts (Melosi, 1981).

The emerging concern about the environment was seen most often in the formation of nature and wildlife groups that would serve as the foundation of environmental activism: the National Audubon Society (1905), National Parks Conservation Association (1919), Izaak Walton League (1922), the Wilderness Society (1935), and the National Wildlife Federation (1936). These organizations would serve as the backbone of the mainstream environmental movement of the latter half of the twentieth century and established the patterns for future political activism on behalf of the environment in the United States. They would also serve as the model for similar efforts in other countries.

From the mid-1920s on, there was a countermovement in the West, composed of ranching and agricultural groups. During the

Stanfield Rebellion (1925–1934), led by Oregon senator Robert Stanfield, western ranchers protested the imposition of fees for livestock grazing in the national forests. The visibility of the movement gave ranching interests a role in decision making that would continue well into the twenty-first century. Under the 1934 Taylor Grazing Act, these interests were granted greater participation in decision making on grazing policies and land management. During the McCarran hearings (1941–1946), initiated by Nevada senator Pat McCarran, a series of public meetings allowed ranchers to testify about the impact of grazing fees and, later, an unsuccessful attempt to abolish the Bureau of Land Management. As had been the case with World War I, World War II seemed to put domestic issues such as the environment on hold.

1945–1960

After World War II, Americans returned to the subjects of pollution and conservation that had, for the most part, been deleted from the political agenda. Though some historians believe the Age of Ecology began on 16 July 1945 with the testing of an atomic bomb in the New Mexico desert, many observers feel that overcrowding in the national park system and periodic episodes of pollution were more important to Americans who had survived the war (Worster, 1977). The Defenders of Wildlife, one of the few major organizations founded during this time (1947), joined the five national organizations established prior to this period, sharing the spotlight and keeping the focus on nature. Keep America Beautiful, established in 1953, came from a different perspective altogether, seeking to encourage Americans to clean up the littered highways that had become crowded with cars after the war. A shortened working week, more opportunities for recreation, and a societal value of leisure brought more Americans into the outdoors than ever before. The more people visited the national parks, monuments, and wilderness areas, the more they began to demand access to them, as well as their preservation for future generations (Hays, 1987).

The 1950s have been described as a time when conservation hardly made a ripple on the public or political agendas because most Americans focused on the Cold War, McCarthyism, and the civil rights movement. Among the exceptions were the Sierra

Club's successful blockage of the construction of Echo Park Dam in Dinosaur National Monument. Perhaps the only issue that captured scholarly attention was whether or not the earth was capable of sustaining population growth—a resurrection of the theories first advanced by Sir Thomas Malthus (Kline, 2000).

1960–1980

There is disagreement among those who have studied environmental activism regarding when the movement "started" or coalesced into a recognizable force in public policymaking in the United States. But there is agreement that one of the major events of the 1960s was the 1962 publication of Rachel Carson's *Silent Spring*. The book, written by a woman with a background in marine biology, warned of the destruction of plants and wildlife by chemical insecticides "showered indiscriminately from the skies" (Carson, 1962, 156). Although she was heavily criticized, the book served to educate the public and provided a scientific foundation for campaigns against pesticides and the substance dichlorodiphenyltrichloroethane (DDT) in particular. The book raced to the top of the *New York Times* best-seller list and won numerous awards.

Prior to the publication of *Silent Spring*, many believed that the country lacked a conservation conscience, owing in part to postwar affluence and a greater reliance upon "the synthetic revolution" of chemicals, plastics, and new fibers. But there was also a change in direction for those seeking a better quality of life rather than a standard of living, resulting in a quest for more leisure time, recreational opportunities, and personal health and security (Sale, 1993). Some of these ideals were reflected in what some call "soft environmentalism," the type of activism that can engage citizens with no political agenda and for which there is little opposition. An example of this form of activity was the highway beautification movement championed by Lady Bird Johnson, the wife of the president. Using southern charm and Americans' desire to clean up the countryside, she encouraged groups to spread wildflower seeds along highways, erect fencing to shield the view of landfills and auto wrecking yards, and stage periodic litter cleanups. The effort seemed genuine and innocuous, but it was blocked by the billboard industry and never really recovered (Floyd and Shedd, 1979).

The major, well-established environmental organizations used this period as a time to reinforce awareness about the need for environmental protection. They focused primarily on the protection of wildlife and public lands with a limited agenda that was understandable and salient to the public. Although some groups began to consider the impact of toxic chemicals and emissions as part of their mission statements, the majority still relied upon the time-honored issues of the conservation movement.

Throughout the 1960s, there was a spurt of membership growth and the development of new organizations. The Group of Ten mainstream organizations met informally to try to build coalitions on major issues, and their membership is estimated to have increased sevenfold, from 123,000 to 819,000 between 1960 and 1969 (Dunlap and Mertig, 1992). Smaller groups with fewer resources and a more recent history tended to have a local support base rather than a national one. Some endured, but others faded as quickly as they sprung up.

Earth Day 1970 is noteworthy as the first (and largest) worldwide environmental demonstration, even though the major groups distanced themselves from the events. But over the next ten years, almost every organization would benefit from the development of the environmental movement and the change in direction from a nature-based to pollution-based agenda. The top five mainstream groups—Sierra Club, National Wildlife Federation, Audubon Society, Wilderness Society, and Izaak Walton League—would increase their membership from 841,000 in 1970 to 1,485,000 in 1980, a change of more than 76 percent. The Audubon Society registered the most new members, from 120,000 in 1970 to 400,000 by 1980, a threefold increase (Sale, 1993).

Other notable developments during this period are covered in greater detail later in this chapter. They include the advent of the green party movement, the emergence of ecofeminism and deep ecology as a philosophy, Arne Naess's distinctions between shallow and deep ecology, and the rumblings of what would later be called ecoterrorism or ecotage as described in Edward Abbey's *The Monkey Wrench Gang* (1975).

The founding of the Environmental Defense Fund (EDF) in 1967 opened the door to a new type of activist organization and approach—the use of litigation by public interest law firms dedicated to protecting environmental quality and public health. The EDF had originally focused on support for a ban on the chemical DDT but expanded its client base by adding scientists to its staff.

The Natural Resources Defense Council (NRDC), in contrast, was founded by lawyers and run by lawyers, although both groups received grants from various foundations to fund their efforts.

Although environmental protection groups dominated litigation throughout most of the 1960s and 1970, opposition from industry interests and some segments of the public was represented by groups such as the Mountain States Legal Foundation (MSLF), founded in 1977. Despite the common belief that business has always opposed environmental protection initiatives, many industries did not really begin to take the environment as a political issue seriously until well into the 1970s. At that point, opposition was usually funneled through industry trade groups such as the Chemical Manufacturers Association or the National Association of Manufacturers (Switzer, 1997).

1980–1988

Although the decades of the 1960s and 1970s are noteworthy for the rise of national and grassroots organizations seeking to protect the environment, the agenda of the 1980s was largely controlled by industry opposition to pollution-control mandates and opposition to federal land use policies. The Sagebrush Rebellion (1979–1983) marked a strengthening of opposition to federal land policies in the West, led by groups such as LASER (League for the Advancement of States' Equal Rights). These groups served as an opposing force to environmental organizations, and the Sagebrush Rebellion, although short lived, also was the foundation for the wise use, county supremacy, and property rights movements that surfaced toward the end of the 1980s and into the 1990s (Switzer, 1997).

Wise use advocates questioned whether decisions about natural resource management on public lands should be made at the federal, state, or local level. They focused their energies on the 1964 Wilderness Act and the 1976 Federal Land Policy and Management Act, which activists claimed were in violation of the "public" nature of public lands. The implied trust language of the property clause of the U.S. Constitution was used to promote the concept that lands in the public domain ought to be disposed of, preferably given to counties. By using the Constitution as the basis for their arguments, the Sagebrush rebels gained a considerable following among western ranchers and farmers, if not regional statehouses (Cawley, 1993).

The legitimization of the property rights movement came about through the publication of books such as Richard Epstein's *Takings: Private Property and the Power of Eminent Domain* (1985). Epstein brought academic credibility to activists who, to some, seemed guilty of trying to invent new rights in their own self-interest. The environmental opposition was most effective in coalition building with events such as the August 1988 Multiple Use Strategy Conference. By piggybacking onto larger organizations, smaller grassroots groups and even individuals created what appeared to some to be a new social movement.

There were two distinctive developments in activism during this period. On the one hand, the national mainstream groups became more professionalized and, in one observer's view, "a relatively harmless and effete new club" that suffered from overblown egos and Potomac fever (Dowie, 1995, 74). The Group of Ten was criticized for being white and male dominated and for limiting its membership by excluding some of the largest and most powerful activist groups, such as Greenpeace. Washington, D.C., became the focus of mainstream activists; the number of registered environmental lobbyists there rose from two in 1969 to eighty-eight in 1985. The number of paid staff members also increased dramatically. By 1986, the Audubon Society had over 200 Washington employees, and the National Wildlife Federation had more than 500. Friends of the Earth moved its operations from San Francisco to Washington in an acrimonious split that some felt drew the center of activism away from the needs of the West. The leadership of groups also changed from the original grassroots activists to professionals familiar with the Beltway: lawyers, accountants, lobbyists, and managers (Sale, 1993).

On the other hand, distancing themselves in Washington offices far from the environmental front, the major organizations were unintentionally responsible for the development of local, grassroots groups whose primary interests concerned the health of their communities. With a smaller membership base, local groups turned to problems that affected their specific neighborhood, rather than the nation. Issues such as the siting of a toxic waste facility, the operation of a municipal waste dump, and clusters of illnesses were much more likely to arouse their constituency than global problems that seemed impossible to resolve. Grassroots groups also became popular because they allowed residents to participate directly in the decision-making process rather than relying upon a distant and often invisible board of directors. Funds raised locally were used locally, and most groups never felt the need to hire staff or open full-time offices. Their members did not need to pay dues, and they met in living rooms or community centers. The emphasis on the local, rather than a larger scale, provided a sense of empowerment that many of the mainstream groups could not provide.

An awareness of the global nature of environmental problems began to grow in the last half of the decade, fueled by a National Forum on Biodiversity (1986) in Washington, D.C. The conference, attended by more than 14,000 people and sponsored by the National Academy of Sciences and the Smithsonian Institution, brought activists together at a time when the Reagan administration had extinguished the fire and passion of many environmental groups.

Prior to this period, environmental activism was associated with white, middle- to upper-class citizens, who had the time and resources to devote to organizations and issues. That image changed in 1982, when the environmental justice movement burst onto the political scene, bolstered by protests by primarily poor, minority residents of Warren County, North Carolina. A year later, a General Accounting Office study on the siting of hazardous waste dumps underscored previously anecdotal reports that linked such land uses to minority communities. But the most powerful evidence came from a 1987 report by the United Church of Christ, which had founded the Citizens' Clearinghouse for Hazardous Waste in 1981. The report reinforced anecdotal evidence that hazardous waste sites were being developed in neighborhoods that were predominantly poor and, often, in residential areas that were home to people of color.

1988–

The end of the administration of George Bush in 1992 marked a watershed in environmental activism, as groups openly cultivated support for the new administration of William Jefferson Clinton. When Clinton named former Tennessee senator Al Gore Jr. as his vice president, many activists believed the president was signaling a new era of environmental protection. Prior to his selection as vice president, Gore had developed a reputation for environmental awareness (a trait Clinton lacked) that leaders believed would become one of Gore's major responsibilities in the new administration. Gore's supporters seemed to overlook the fact that as governor, Clinton's environmental record in Arkansas would have been unacceptable under any other circumstances. The reality was that the general public was virtually uninterested in the environment—one survey of voters found that fewer than 10 percent of the voters said that environmental issues were important in their decision on who to vote for (Sale, 1993).

But the Reagan and Bush administrations had promulgated regulations that so favored industry and business interests that environmental groups struggled to keep up. Early Clinton administration initiatives were focused on the presence of gays in the military and on national health care reform, not the Endangered Species Act or Superfund reauthorization. As a result, there was a reduction in national organization administration; Greenpeace laid off workers and closed several offices in 1997. "Green" legislation seemed to stall in Congress—bills opposing clear-cutting in forests and plans to drill for oil in the Arctic National Wildlife Refuge went nowhere. The most that could be hoped for was damage control—preventing statutory rollbacks—and the possibility that groups could come to some agreement about what issues to prioritize. The Sierra Club even tried to stay out of some of the more contentious debates so that it could maintain its "reasonable" reputation in comparison to the more hostile or militant organizations.

Although environmental groups seemed caught up in disputes and a lack of a consensus agenda, the wise use, county supremacy, and property rights groups began to prosper. In May 1989, wise use advocates rallied in Salem, Oregon, to protest logging bans on the habitat of the northern spotted owl. Their simplistic message—"jobs versus owls"—resonated with many Americans who felt most environmental problems had already been solved. Some believed that activists were now going too far, sacrificing the U.S. economy for a small bird most had never heard of before.

As the new millennium dawned, it became clear that environmental protection lay dormant on the political agenda. A booming economy and concerns about a dreaded computer bug, Y2K, overshadowed any worries about the natural world. Gas prices stabilized and then dropped; more energy efficient cars became widely available; there were more national parks, refuges, wilderness areas, and monuments than ever before; and all seemed well on the environmental front. The 2000 presidential election and questions about hanging chads clouded the decade's first year, and terrorist attacks on September 11, 2001, made virtually any other issue seem insignificant. The national environmental groups found it difficult to do fund-raising when billions of dollars were needed to fund the war effort and rebuild the economy after the attacks on New York and Washington, D.C. Even hinting that there might be other important issues seemed almost unpatriotic. American flags replaced decals for the Nature Conservancy on activists' gas-guzzling sports utility vehicles.

Although the administration of George W. Bush seemed to put most domestic policy issues on hold, there were some attempts to revert back to a more conservative environmental agenda—a fact that did not go unnoticed by mainstream groups who accused the president of sneaking proposals through the back door. Efforts to modify the ambitious Clinton administration forest proposals, plans and support from Interior Secretary Gale Norton to open up the Alaska National Wildlife Refuge to oil drilling, and continued debate on the expansion of the national monument system put groups such as the Sierra Club and Wilderness Society on high alert. They aimed their efforts at Washington and supportive members of Congress, especially the Democratic-controlled Senate, to make sure that no major policy changes or initiatives escaped their scrutiny while the rest of the nation was consumed by war in Afghanistan, political unrest in Argentina, and a hoped-for economic recovery.

But even though there were few changes in environmental legislation per se, the Bush administration did work hard to reduce the impact of the Clinton years by overhauling various environmental regulations. Wetlands, for instance, which had been the focus of considerable activism during the 1980s, were given less protection when the president's staff announced a rollback in the permits needed to fill in ditches and other seasonally wet land. Other targets were drinking water standards and the Endangered Species Act. Some proposals took the form of trial balloons—released by the White House as a gauge of public reaction and then usually abandoned.

The new century began with a political agenda that was so filled with nonenvironmental concerns that groups struggled just to be heard above the din of other voices. The questions about whether or not the economy was poised for recovery, or whether defense spending would grow at the expense of social problems, forced the environment off most decision makers' agendas altogether.

Types of Activist Groups

The Mainstream Organizations

By "mainstream," most researchers refer to environmental organizations that use traditional political tactics such as lobbying members of Congress, initiating letter-writing campaigns, or mobilizing their members through local, state, and national meetings. Their goal is to participate in the drafting of legislation or to influence a member of Congress. Variously, the mainstream groups are described as the Group of Ten or Group of Twelve. This includes the three Progressive Era groups (Sierra Club, National Audubon Society, and National Parks Conservation Association), the three "between the wars" organizations (Izaak Walton League, the Wilderness Society, and the National Wildlife Federation), the post–World War II Defenders of Wildlife, and the five major groups that formed in the late 1960s and early 1970s: Environmental Defense Fund, Friends of the Earth (which incorporated the Environmental Policy Institute), Natural Resources Defense Council, Environmental Action, and Greenpeace Action, the lobbying arm of Greenpeace USA. They are distinguished from other national groups because of their lobbying activities, and they represent the most important sources of influence and membership (Mitchell, Mertig, and Dunlap, 1992).

These groups are enduring, and they have an established membership base, expertise, financial resources, and a track record of success in Washington. Many of the mainstream organizations are based in Washington, D.C., with state-level or local chapters or branches. Some of the smaller chapters operate almost independently of the national organization, using the group's name and reputation to attract members who are often more interested in local or regional issues.

Species- and Issue-Specific Organizations

One of the features that has characterized the past two decades of environmental activism has been the increasing number of issue- and species-specific organizations. Although most of the major national groups have broadened their issue agendas, with an increased emphasis on global problems, there has been a corresponding growth in smaller groups with a more narrow focus. It seems as though virtually every species now has its own protective organization, from Bat Conservation International to groups protecting white-tailed deer, wild turkeys, and manatees. Other groups focus exclusively on pesticides, coastlines, or the dangers of radiation.

The major problem with this type of organization is that groups often must battle among themselves for very scarce resources, and there is little, if any, attempt to build coalitions. Issue-specific groups, for instance, may have difficulty justifying cooperative relationships with any other organization whose environmental agenda does not match their own. Turf battles to declare who "speaks" for the majority on an issue, such as wildlife and habitat protection, occur even when groups want to accomplish the same basic goals. Species-specific groups are loathe to share contacts, memberships lists, or political forums with any other species group. At times, groups that are in support of protection for a single animal, such as elephants, may be competing with a half dozen others who believe that "they" are the voice for the species. The lack of cooperative efforts fragments membership and, therefore, activism.

Environmental Justice Groups

Environmental justice activists have tried to do two things. One is bringing attention to examples that recount the efforts of communities and local governments that have been "shortchanged" by the political, social, and economic systems. The other is attempting to quantify how much racial injustice there really is in the United States, showing inequalities in the distribution of environmental burdens and benefits. The groups are said to have evolved from the ecocentric perspective of mainstream organizations (the protection of nature, global warming) to one that has been termed homocentric, or human centered. Environmental justice activists seek to bring attention to concerns about human health (especially that of children exposed to toxic substances), food safety, urban pollution and the siting of waste facilities, and working conditions for agricultural workers. They argue that the mainstream groups have failed to take the interests of minority groups into account, as exemplified by only token representation on the boards of directors of the groups. There is a divisiveness among some of the environmental justice advocates, however, who are at odds as to whether their best strategy is to piggyback on the organization and resources of mainstream groups or to work toward resolution of their own specific interests (Melosi, 2000).

Critics believe that the current environmental justice debate will have little effect on federal policymaking and may, in fact, exacerbate existing problems. The empirical evidence that indicates disproportionate siting or undue economic burdens is actually weaker than some advocacy groups have stated. Without scientific justification for their claims, there is a belief that Congress is unlikely to yield to major policy changes. In addition, there are considerable limitations and drawbacks of a focus on environmental justice that fails to define and pursue a coherent set of environmental policy priorities or of an unwillingness to face "politically inconvenient facts" about environmental health risks (Foreman, 1998, 4).

Still, grassroots environmental activism against environmental racism, as it is sometimes called, is alive and mostly well, although highly localized. Groups such as West Harlem Environmental Action have challenged the siting of a water pollution control plant; Chester Residents Concerned for Quality Living brought a lawsuit in Pennsylvania that received considerable media coverage and attracted university students who formed the Campus Coalition Concerning Contamination; representatives of the Gulf Coast Tenants Association in Louisiana have testified before Congress about the industrial corridor they call "Cancer Alley."

Green Political Parties

Green political parties have attempted to create a national dialogue on the electoral front, including the support of presidential candidates. They first made themselves a visible political presence in the United States in 1984, and most of their activism has come at the state level. They formed loosely knit groups on a regional level to form national Interregional Committees, later changing their name to the Green Committees of Correspondence, a reference to the American revolutionary groups. The Interregional Committee was the major decision-making body in the early years of the Green Party, which had its first gathering in Amherst, Massachusetts, in 1987.

On a positive note, Greens have been lauded for their openness, inclusivity, and democracy. Discussions are open ended, participation is encouraged, and debate can be fiery. The flip side of their activism is that they often suffer from what one writer calls "the tyranny of structurelessness"—situations where there is so much "democracy" that arguments become heated and hostile, individuals' feelings are deeply hurt, and the group becomes paralyzed (Rensenbrink, 1999, 5).

Because of the Greens' limited resources and fragmentation, their only real electoral successes have come at the local and state levels. Some city councils (such as that of Arcata, California) have been "greened" to the point where a majority express support for environmental issues. Linda Miller, who ran for the U.S. Senate from Hawaii, received an impressive 14 percent of the vote; Carol Miller, a candidate for the Third Congressional District of New Mexico, received 17 percent of the vote in 1997. Audie Bock won election to the California legislature from a previously "safe" Democratic district.

The presence of the Greens as a viable third party, while focusing attention on environmental issues, has not come without cost. Jonathan Carter, for instance, ran for Congress in 1992 in Maine's Second District and picked up over 10 percent of the votes cast. But he was heavily criticized because Democratic leaders believed his candidacy had enabled Republican Olympia Snowe to defeat the Democratic candidate; Snowe used her congressional seat as a platform for a successful run for the U.S. Senate. In 2000, a backlash developed against the Green Party movement in the United States when presidential candidate Ralph Nader was said to have siphoned off votes for the Democratic Party candidate, Al Gore, leading to a very narrow victory by Republican Party candidate, George W. Bush. Had Nader not entered the race, some believe, the outcome would have been totally different. Green Party leaders, however, said that Nader's candidacy was the only way to ensure that the environment was placed on the presidential election discussion agenda.

The successful state elections have allowed the Green Party to become officially recognized as a political party. In 1994, Carter's subsequent run for governor made Maine the first state east of the Mississippi to give ballot access to the Greens side by side with the Democrats and Republicans. This tactic would give Green candidates the necessary political platform they would need to become part of the environmental political debate. But in a political system that does not allow for proportional representation, such victories in the United States have been few and far between.

Spiritual Environmentalism/Ecotheology

There is a quiet form of activism, known as spiritual environmentalism, or ecotheology, that has attracted a totally different kind of follower. It appears to have had its roots in the Faith-Man-Nature Group (1964) sponsored by the World Council of Churches to explore the relationship between nature and humanity. Twenty years later, the North American Conference on Christianity and Ecology was established to assist theologians in developing a worldview of Christianity and nature.

The most recent manifestation of the movement can be traced to 1992, when Calvin DeWitt founded the AuSable Institute of Environmental Science in Michigan as a way of combining biology and ecology courses with scriptural references. He joined with author Anthony Campolo in 1993 to establish the Evangelical Environmental Network, which serves as the conservative Christian representative in the Religious Partnership for the Environment, made up of Jewish, Catholic, and other religious groups. Another national group is the Christian Society of the Green Cross based in Pennsylvania (Smith, 1997).

Although there are few traditional membership organizations within this segment of the movement, there has been the development of a Christian environmental ethic—a moral commandment to watch over all creation—and alternative views of creation that incorporate religion and science. The ethic has been reinforced by several religious leaders, including a 1989 papal encyclical from Pope John Paul II and a 1991 Statement by Religious Leaders at the Summit on the Environment that brought together twenty-four major U.S. religious groups. The call to action is particularly strong in the western United States, built in part on financial support from environmental grant-making organizations.

The primary issue that has united such diverse groups is the Endangered Species Act, which has been threatened for revision by leading Republicans in Congress. Groups such as Green Cross have met with political leaders, engaged in lobbying, and started a "rescue God's Creatures" campaign. But supporters have not been able to convince media-savvy religious leaders to join their cause, although noted evangelical Billy Graham did participate in a parade and noted his own deep and increasing concern for the environment (Smith, 1997).

To date, spiritual environmentalism has not inspired massive grassroots activism. Much of the recruitment of potential members is done by supporters who visit churches and attempt to engage potential followers at the local level. They admit that theirs is a long road and that not all religious leaders agree with their message. But they use the Bible as their guide and prayer as their most powerful form of activism.

Deep Ecology

During the 1970s, the writings of Norwegian philosopher Arne Naess brought together a different group of activists, who used the term *deep ecology* to describe their view of the relationship between humanity and nature. Naess had been influenced by the work of Rachel Carson and the first rumblings of environmental activism in Norway. A mountaineer as well as a philosopher, Naess argued that there were differences between what he called "shallow" and "deep" ecology. Shallow ecology was seen as a short-term perspective that dealt primarily with such issues as pollution and the use of natural resources, whereas deep ecology sought to realign both worldview and lifestyle to take resource depletion into account. Deep ecology also had a scientific basis— a growing awareness of interconnectedness and the development of knowledge about ecosystems.

George Sessions and Bill Devall, who brought deep ecology to the United States, contrasted this worldview with mainstream environmentalism by stressing the interrelationships between humans and nature but also granting species the right to survival. Those who supported Naess's view also sought to preserve the diversity of all species, the reduction of human consumption, and the preservation of wilderness. American poet Gary Snyder added his commentary on the environmental practices of Native Americans and Zen Buddhism, producing various hybrid philosophies.

There are numerous critics of deep ecology, primarily among philosophers who argue that the concept ignores social issues. Murray Bookchin, for instance, noted that Naess and his followers supported population reduction but did not relate policies to issues of justice and equality. Some radical activists accepted deep ecology as a reason for militancy, especially when nations such as the United States engaged in what they believed was unnecessary resource consumption and exploitation of species (Devall and Sessions, 1985).

Deep ecology has produced a broad foundation in literature written by scholars from around the world, but it has not become a "movement" in the political sense. Those who believe in its tenets may become involved in direct action, such as protests in Seattle against the World Trade Organization, but there are no formal groups that engage in promoting its views as part of a political strategy.

Ecofeminist Organizations

Around the early 1970s, some scholars began to argue that the environment is a feminist issue, drawing connections between women and nature. This philosophical perspective, ecological feminism, or ecofeminism, has several themes: that our current environmental crisis is the result of patriarchal culture; that women and children bear a disproportionate share of the health risks and problems caused by pollution; that the abuse of nature is akin to men's abuse of women; and that there is a spiritual connection between women and nature expressed symbolically by the goddess theme or Mother Nature (Merchant, 1980; Warren, 1993).

Interestingly, ecofeminism has had more of a philosophical influence on other organizations that embrace it, rather than leading to the formation of activist groups. It has also expressed itself more in global organizations rather than in an activist form in the United States. The *chipko* movement in India, the Brazilian Acao Feminea Democratica Gaucha, and women's rural progress associations in Zimbabwe are responses to previously male-dominated environmental and agricultural movements in other countries.

Like the green party movement, most ecofeminist organizations are inclusivist and often involve Native Americans or indigenous peoples. Groups may focus on values such as care, love, friendship, trust, and appropriate reciprocity—relationships with an "other" that seem distant from traditional political activists. Thus, the ecofeminist organizations are less likely to attempt to become involved in mainstream political activities such as lobbying or letter-writing campaigns. Their influence on environmental politics is much more subtle, although some groups do engage in direct action protests. In general, however, ecofeminist activists have agreed to convene informally, if at all, and some of their underlying tenets are gradually being absorbed by some of the global organizations.

One of the exceptions would be the activities of Concerned Citizens of South Central Los Angeles, a group that came together in 1986 to fight against a waste incinerator that had been proposed for their neighborhood. Although the group was open to both men and women, it became clear that women would dominate the discussion and the group's tactics. Some observers believe this was because the issue raised their community consciousness and because they perceived a threat to their homes and families. Even though the majority were poor and undereducated, they were joined by other women from the Los Angeles area who were politically aware, who had resources of both time and money, and who crossed racial lines to join in with local residents.

Direct participatory action in crises or in cases where there is a perception of oppression against women may result in activism, but this is most likely to be short term, lasting only as long as the problem is unsolved (Hamilton, 1993).

Foundations

Richard Goldman, chief executive owner of a major independent insurance brokerage, and his wife, Rhoda, who is a descendant of Levi Strauss and the apparel company he founded, support activists through their own charitable organization. The Goldman Environmental Foundation was founded in 1989 to recognize the efforts of individuals to preserve or enhance the environment. Recipients, who are announced in April of each year, are given a cash prize to enable them to continue their advocacy efforts. Since the awarding of the first prizes to six grassroots activists in 1990, award recipients have come from countries all over the world. They are nominated by a network of international environmental organizations and a confidential panel of citizen activists and prominent policymakers.

In 1990, the six recipients included Lois Gibbs, a U.S. activist in her community of Love Canal; Janet Patricia Gibson, a biologist whose work has helped preserve the fragile barrier reef off Belize; Janos Vargha, a writer who opposed the Hungarian government's plan to dam the Danube River; Harrison Ngau, a leader of an indigenous peoples group in Borneo that is fighting logging operations in the rain forest there; Michael Werikhe, the "Rhino Man" of Africa, who has led the efforts to save the black rhinoceros; and Bob Brown, whose successful campaign to preserve the Franklin River brought environmental protection to the attention of the people of Australia. Since then, activists on each of the earth's continental regions have been recognized with the prestigious award.

Other foundations, such as the one established by media mogul Ted Turner, are designed primarily to finance research projects or educational outreach. Because they have tax-exempt status as charitable organizations, they rarely engage in any type of political activism. But the Goldman Prize is considered by environmental activists as being in the same category as the Nobel Prize, and the winners receive considerable media attention for their cause. The foundations serve an important purpose by increasing visibility and providing financial resources for activists and organizations that sometimes otherwise would be ignored or financially unstable.

Strategies and Forms of Environmental Activism

From the beginning, mainstream environmental activism followed a typical pattern of political participation that had become the hallmark of labor unions, social welfare groups, and, to a lesser extent, business interests. Typically, groups would coalesce around an issue, conduct a letter-writing campaign to their elected representatives, court the media, send a handful of members to Washington to directly lobby lawmakers, and later hire professional lobbyists to represent them. The groups with the most resources, such as the National Audubon Society, opened offices in Washington or New York as a way of establishing a national presence. These tactics may have hurt the groups, however, as they began to be perceived as hierarchical and top heavy with administrators, diluting their message (Sale, 1993).

The environmental groups' activism differed somewhat from that of other social movements, such as advocates for civil rights, who were much more likely to engage in direct action protests than in drafting legislation or meeting with congressional staff members. This is in part due to the difference in resources. Many of the mainstream organizations were able to rely upon the expertise and political acumen of activists from supporters of the women's movement who had worked toward passage of the Equal Rights Amendment. They tended to have access to policymakers, resources, and long-term issue agendas that lent themselves to national legislation. The following sections outline their most common political strategies.

Monitoring Voting Records

The legislative process is complex and often difficult to follow, even for those who are within Washington, D.C., and familiar with its intricacies. Most activists, however, do not live in the nation's capital and rely upon others for information about emerging legislative initiatives and new laws. It becomes almost impossible for any one individual or interest group to stay on top of the fast pace that successful legislative monitoring requires. Several groups have responded by providing their members with information about a legislator's voting records, especially on key environmental issues, which can then be communicated to the public. Sometimes this type of information can be useful as interest group members make decisions on which candidate to support or oppose. When the legislative scorecards are reported by the media, groups also hope to influence public attitudes for or against particular members of Congress.

The League of Conservation Voters (LCV), founded by Marion Edey in 1970, has become known as an authoritative group that identifies environmental voting habits and communicates that information to its constituency. It has been called "the political arm of the environmental movement" (League of Conservation Voters, 1996, 5), and the group refers to its "electoral muscle" (Annual Report, 1996, 3) on behalf of the environment. When LCV issued its first *National Environmental Scorecard* in 1970, it set the stage for its annual accounting of "how green, or not so green, our federal elected officials' voting records are" (League of Conservation Voters, 1996, 18).

The group's political committee meets regularly to examine the votes recently cast and determines which ones are the most significant (but which also present members of Congress with tough choices), consulting with more than twenty-five other groups and organizations. They then research the congressional actions and make a recommendation on a legislator's "score" on that vote. This process allows for considerable input and reflects the environmental community as a whole, rather than a single organization. Once the scores are tallied, LCV holds a news conference in Washington, D.C., and organizes national releases of the information that same week. The scorecard is also available to the public on-line. For the final session of the 106th Congress, for example, members were evaluated for their votes on seven Senate and fourteen House issues and given a composite score ranging from zero to 100. In the Senate, the Rhode Island delegation received a perfect score (100), with the four Alabama and Alaska senators receiving scores of zero. In the House, Vermont scored 100, with Idaho the lowest with 2 percent. Individual representatives also ran the spectrum from zero to 100 (League of Conservation Voters, 2000).

Although the LCV is engaged in political action by supporting and opposing candidates, it maintains its status as a nonprofit organization (making contributions tax deductible) by engaging in activities that are not subject to federal contribution guidelines, such as media programs and voter education. It also has a political action committee, the LCV Action Fund, which is subject to yearly campaign contribution limits. It has served as a model for other groups who prepare similar scorecards, such as the Sierra Club, Zero Population Growth, and the environmental opposition organization, the American Lands Rights coalition.

Initiatives and Referenda

Another way in which environmental activists have sought to directly influence the political process is through citizen initiatives, unique to some state constitutions. By collecting a sufficient number of signatures, groups can place an issue on the ballot at the local or state level. Some observers believe that the initiative process circumvents the legislative process, which is designed to be more representative of constituents. This has become especially true in states where signature gatherers are not volunteers, but individuals paid for each valid signature they gather even though they may have no interest in the outcome of the vote.

Nowhere is this more apparent than in California, where citizen initiatives have become commonplace. In November 1990, environmental groups gathered sufficient signatures to place Proposition 128 on the ballot. The initiative, nicknamed Big Green, contained a 16,000-word list of thirty different measures on environmental issues, including those that would have banned a variety of cancer-causing pesticides, prohibited new offshore drilling, and stopped the cutting of redwood trees. At the same time, another measure, Forests Forever, would have banned clear-cutting of old-growth forests.

To counter Big Green, a loosely knit coalition of industry groups and interests placed two other measures on the ballot. The Global Warming and Clear-Cutting Reduction, Wildlife Protection, and Reforestation Act, termed Big Stump, was coupled with another industry-sponsored initiative, the Consumer Pesticide Enforcement Act, known as Big Brown. These initiatives joined a state ballot that was already crowded with twenty-eight measures that voters had to sort through. Millions of dollars were spent to support and defeat the initiatives, as voters became more and more confused by too much information on too many subjects. The initiative failed to pass. Similarly, a 1996 ballot initiative was brought by Californians for Balanced Wildlife Management, a group actually composed of hunters seeking to legally kill mountain lions. The hunting group was opposed by the California Wildlife Preservation Coalition, with both groups trying to educate the public (Switzer, 1997).

It has been argued that one of the reasons why Al Gore lost the presidential campaign of 2000 to George W. Bush is that there were no major state-level environmental issues on the ballot. The absence of citizen initiatives or legislative referenda, which traditionally have been responsible for increased turnout, may not have been the most important factor in the election. But if there had been key issues placed before the voters, it is possible that individuals who otherwise might not have voted would have gone to the polls.

Providing Expertise and Information

It is commonly believed that environmental organizations are at a political disadvantage when compared to industry interests because of the tremendous disparity in resources. But in the past thirty years, activists have developed a sophisticated information-gathering and -dissemination scheme that equates with almost anything a business sector can produce. Activist organizations have recognized the need to provide their own experts at congressional hearings, to engage committee and subcommittee staff members as legislation is being drafted and amended, and to put their members on legislative alert when constituent support for or against a measure can be mobilized.

One of the least well known but sometimes most effective tactics used by activists involves participation in the administrative implementation of statutes, or rule making. Once Congress passes and the president signs legislation, agencies such as the U.S. Forest Service or the Bureau of Land Management must take steps to put the law into motion. Rule making grew out of the Administrative Procedure Act of 1946 as a way of bringing predictability to the decision-making process of government. Rules are sometimes called "the products of the bureaucratic institutions to which we entrust the implementation, management, and administration of our law and public policy" (Kerwin, 1999, 3).

Rule making involves the aggressive monitoring not only of the laws themselves but also of subsequent actions by agencies. Rule-making announcements appear in the government's *Federal Register,* inviting the public to participate and comment. The process is structured with requirements related to public comment periods and notification, but there are thousands of new rules proposed each year. Most rule-making sessions are held outside the scrutiny of the media or the public, not because they are secret, but because most people are unaware of the process and how it affects them. The scope of rules may be narrow or broad, depending upon the policy being considered, and the impact can be significant. During the 1970s, at the same time the environmental movement was at its strongest and major environmental statutes were being enacted, Congress was also delegating much of its authority to the bureaucracy—the unelected part of the executive branch most often responsible for rule making.

The passage of the National Environmental Policy Act, the Clean Air Act, and the Clean Water Act, for instance, established the need for an extensive broadening of the Environmental Protection Agency's power. The agency's staff was often given unrealistic deadlines for preparing guidelines on how an environmental impact statement was to be prepared or ways to get polluters to comply with the new laws. It is here that environmental groups often play a key role. Agencies overwhelmed with new legislation, and understaffed with technicians and scientists, are often eager to rely upon the information provided by those external to the bureaucracy.

Groups are able to provide various types of substantive information about the impact of a statute on their members, industry sector, or the economy. During the rule-making phase of the 1990 Clean Air Act, for example, groups such as the American Lung Association were active in testifying at hearings about the health impact of criteria pollutants, especially on young children. At the same time, representatives from power companies monitored rules that would require them to install costly pollution-control devices. They also testified at rule-making hearings about whether new requirements were technically feasible and at what cost for what benefit. Some observers believe that rule making is when agencies act like legislatures (Shapiro, 1986).

One strategy that has always had mixed results, regardless of the issue, is the circulation of petitions. These informal documents, which in reality have no legal basis, are often generated by grassroots groups to show policymakers that there is public support for or against an issue. Typical is the case of Anti Oahe, a small group of farmers in Brown County, South Dakota. During the 1950s and 1960s, farmers had opposed a U.S. Bureau of Reclamation project to provide agricultural irrigation to an area in the James River Valley that had been dry-farmed for decades. The project was also the most expensive federal project ever developed in the state.

Even though the government argued that the project would provide economic benefits such as the stabilization and expansion of agricultural production and income, as well as 3,000 new jobs, Anti Oahe began circulating petitions among the local residents. The group sent one of its representatives to Washington, D.C., with the petitions, expecting that elected officials would jump to attention and see the power of the opposition. But the representative was totally ignored, the petitions having no effect on government officials (Carrels, 1999).

Conducting Public Relations Campaigns and Outreach

In 2001, the National Environmental Trust published a report, *Permit to Pollute,* that uncovered deficiencies, backlogs, and oversights in Ohio's Environmental Protection Agency, which issues operating permits to factories and power plants within the state. Such reports are often lost in the thousands of press releases sent daily to the media or are read one day and forgotten the next. But in this instance, a coalition of environmental groups went a step further by launching a monthlong, $20,000 advertising campaign of billboards, radio spots, and newspaper displays. They urged Ohio governor Bob Taft to enforce the permit program, calling him soft on environmental protection. Even though the coalition could only afford a sustained advertising barrage for one month, it gave added clout to the report and put the governor's office on the defensive (Smyth, 2001).

Even smaller grassroots groups turn to commercial outreach. The Moab, Utah–based Glen Canyon Action Network seeks to force the government to dismantle the Glen Canyon Dam and drain Lake Powell. In addition to the usual rallies, the network candidly admits that it will measure its success by how much publicity it gets. One of the ways the network publicizes its goals is through the Restoration Creamery, an ice-cream parlor that serves up pamphlets dealing with Colorado River restoration and pointedly named flavors. There is Abbey's Rocky Road (a reference to Edward Abbey, the author who supported the dam's demolition) and Music Temple Almond Fudge, which is named after a now-submerged grotto under Lake Powell. The group uses the profits from the creamery for its efforts, and the storefront operation provides visibility.

There is also an opposite marketing strategy that has proved effective for many organizations—boycotts and bans. Among the initial efforts was one started by the United Farm Workers (UFW) in 1963 against farmers growing California table grapes. Led by noted activist Cesar Chavez, the boycott became a focal point for activists seeking to improve the labor conditions of farmworkers. The boycott was in effect for sixteen years before UFW leaders decided that many of their concerns had been met ("New Initiatives," 2001).

Twenty years later, boycotts against Starkist Tuna affected the entire industry. The Earth Island Institute alleged that the company had killed too many dolphins in large nets designed to catch the tuna, and a nationwide boycott began. Within three years, the company and virtually all others began to label their products as "dolphin safe" as they stopped the practice of using drift nets. Other boycotts were initiated against Exxon gas stations after the 1989 *Exxon Valdez* oil spill in Alaska and against the Nestlé Company for its involvement in the marketing of infant formula to developing countries (Coolidge, 1996).

Boycotts need not be organized on a national level to be effective, however. Natural food markets in Texas and Colorado have announced that they will not sell genetically engineered products—what they call Frankenfoods—in their stores. Critics of the products believe there are unanswered questions about the safety of the products, both from a health and an environmental perspective. Activists can track boycotts through publications such as the Seattle-based *Boycott Quarterly* and Co-Op America's *Boycott Action News.*

Attempts to educate the public about environmental hazards have also proved to be effective. In 1986, a coalition of groups called the Pesticide Action Network began its Dirty Dozen campaign to focus public attention on the pesticides considered to be the most damaging. The coalition's goal is to continuously educate policymakers and citizens about the use of pesticides and how the same effect can be achieved through the use of sustainable, ecologically sound alternatives. Its efforts have led to bans or severe restrictions on these compounds throughout the world. By utilizing a relatively simplistic list technique, there was less of a need to provide the public with complicated information about chemicals. Instead, the practice was similar to the Ten Most Wanted lists of law enforcement agencies, with the connotation that this campaign would put people on alert for pesticide use in their area.

One of the more unusual public relations tactics has been the use of an "environmental air force" that provides lawmakers and journalists with a bird's-eye view of environmental degradation. The nonprofit group LightHawk uses its 150 volunteers and pilots and planes to fly nearly 1,300 passengers a year over areas of urban sprawl, oil and gas fields in Alaska, and clear-cut forests (Clarren, 2000). Since 1979, the group has lent its services to provide visitors with views of sites that they otherwise would not be able to understand from the ground.

Membership Recruitment

Most of the mainstream organizations saw tremendous increases in membership during the 1970s and 1990s. For example, the National Audubon Society's membership increased from 32,000 in 1960 to 148,000 in 1970, and to 400,000 by 1980. Similarly, the Sierra Club rose from 180,000 in 1980 to more than 600,000 by 1990; the Wilderness Society grew from 45,000 members in 1980 to 350,000 in 1990 (Kraft, 2000). The success of mass mailings, a tactic that became common during the 1980s, created a generation of "checkbook activists" who were members only on paper but who rarely took any active role in the organization's activities. The large mainstream groups relied upon members not only for their donations but also for an indication of their political clout in sheer numbers. Organization officials often cited their growing membership as evidence of increased public support for environmental protection, and individuals often contributed to more than one group as mailing lists were shared.

The marketing of membership changed somewhat during the 1990s as groups began offering membership premiums that sometimes resulted in only small percentages of real contributions to the organization itself. The Yosemite Fund offered nature postcards; the Sierra Club offered tote bags; and most groups also included a subscription to their monthly publications as an additional incentive. Cars began to sport decals from the Cousteau Society and the Nature Conservancy as these checkbook members wore their environmentalism publicly. Although some members did respond to repeated requests for contributions (especially for tax purposes at the end of the calendar year), many seemed content to read the glossy nature magazines they received that required little or no activism on their part. Others joined groups for "accessory" benefits—discounts on hotels, car rentals, or group-sponsored adventure travel.

Supporting and Endorsing Political Candidates

One of the factors that has limited political activism among some environmental groups is the prohibition against supporting political candidates that is the bane of all nonprofit organizations. In order to maintain their tax-exempt status with the Internal Revenue Service, the groups must separate out their educational activities from those that are clearly political. The League of Conservation Voters, mentioned previously, does this in two ways. First, it helps raise funds for its "environmental champions" who are named to its EarthList—candidates who are pro-environment and engaged in tough electoral contests. EarthList candidates also receive help from groups who lend environmental activists to targeted campaigns. Second, LCV identifies its Dirty Dozen incumbents and challengers whom they seek to defeat. The group actively solicits contributions as direct gifts, bundling contributions to targeted candidates to indicate environmental support and seeking gifts of securities and bequests. Although endorsements are useful to political candidates, the donation of expertise and checks can often make more of a difference in ensuring electoral success. Thus, many groups have now formed political action committees to express their political support or opposition.

Building Coalitions

There are many examples of instances in which environmental activists have banded together to form coalitions in opposition to or in support of an issue. This often happens when an issue has a national clientele or when the clout of a single group can be magnified by joining together with similar groups. One of the most noteworthy examples of coalition success took place in the late 1950s and early 1960s when the Wilderness Society drafted legislation to create a wilderness protection system. The organization had been unable to interest Congress in the measure, and hearings dragged on for years. But when the Wilderness Society stopped relying upon its own resources and began to network with other groups, they were able to mount a massive public relations campaign that resulted in thousands of letters and telephone calls being sent to members of Congress. Traditional mainstream groups such as the Sierra Club, Audubon Society, and Izaak Walton League worked with nonenvironmentally-oriented organizations such as the National Jaycees and the Federation of Women's Clubs that brought in an entirely new set of grassroots activists. By 1963, the wilderness bill had passed the House and in 1964, the Senate, after which it was signed into law (Sale, 1993).

The groups associated with the environmental opposition have also been extremely effective in coalition building. It is not unusual for dozens of groups to be listed as supporters of an umbrella organization such as the Alliance for America. Sometimes the coalition is temporary to support or oppose a single proposal, such as the Yellowstone Vision of the Future. In 1990–1991, coalition groups such as People for the West! were brought to Montana State University in mining industry-chartered buses to attend rallies opposing the public hearings on the Yellowstone plan.

Corporate interests have had a profound influence on certain pieces of environmental legislation when they have joined together. In 1988, for instance, business interests were being criticized for their inability to match environmental groups' expertise in the debates over the amendments to the 1977 Clean Air Act. Congressman John Dingell of Michigan called together industry lobbyists to form the Clean Air Working Group (CAWG), which had met informally since 1981 to provide legislators with information on acid rain issues. Dingell was concerned because there was no united front among the various trade associations, companies, and general interest organizations such as the Chamber of Commerce and the Motor Vehicles Manufacturers.

CAWG became a formalized group with an executive director, ten teams handling separate issues related to air quality, and a strategy base three blocks from the Senate office buildings. The effort resembled a political campaign, including telephone banks and computer databases. The group's members toured twenty-five cities, holding press conferences and briefing newspaper editorial boards, along with participating in community debates and radio programs. The result was an effective campaign that brought industries together on a bill that would otherwise have totally divided them (Cohen, 1992).

Conducting Protests and Demonstrations

There are those who believe that protests are as American as apple pie, especially if you look back at American history to events such as the Boston Tea Party. But there have been some periods of time when protests were more publicly accepted than others and some issues where protesters piggybacked on another issue to broaden their base of support. The civil rights movement and subsequent demonstrations for the rights of women, gays, and persons with disabilities have set the tone for most contemporary environmental activists. There is a sense among most demonstrators that violence and destruction are antithetical to preserving nature in its many forms, and as a result, nonviolence and passive civil disobedience have characterized the majority of environmental protests.

In September 2000, a group of around thirty people staged a mock funeral in Nevada's capitol with a casket shaped like the state. Their protest was against the dismantling of the state's water-planning agency and the potential loss of native plants and fish. A member of the Friends of Nevada Wilderness dressed in black mourning attire—a common action to emphasize an environmental loss.

Members of the environmental opposition also use protests as a tactic. In Missoula, Montana, hundreds of logging trucks and demonstrators circled the downtown area in June 2000. They attended a barbecue and rally against a proposal by the U.S. Forest Service to protect about 43 million acres of roadless forests. About 2,000 ranchers, loggers, and off-road vehicle enthusiasts from throughout the state gathered, and nearly half sought to give the federal agency their comments at a meeting. Similar to what had happened in Nevada, pallbearers in hard hats carried a pine coffin into the lobby of the hotel where the meeting was being held.

Not all protests are noisy, nor do they involve large groups of demonstrators. The term *bearing witness* comes from nonviolent groups such as the Society of Friends and the Buddhist religious tenets. It is connected to the spiritual elements of environmentalism and is sometimes associated with the concept of speaking for those who have no voices (Grossman, 1988). It might mean standing near the ocean shore, silently watching as a whale carcass washes in and out with the tide. In the redwood region of Westport, California, one young woman protested against logging operations by removing her blouse whenever a timber company truck drove by. Her actions were filmed as part of a documentary called "The Bare Witch Project," and she says the out-front approach allows her to speak for Mother Earth. Even though there may be an element of drama to such expressions of concerns, their symbolism is often very serious for the participants.

Ecoterrorism

Author Edward Abbey is credited with inspiring radical environmental groups through his novel, *The Monkey Wrench Gang*, published in 1975. Some environmental activists used the book as a guide for ecoterrorism (for details, see "Does Ecoterrorism Work?" in Chapter 2). The tactics of the groups vary, but they focus on civil disobedience, sometimes violent, to thwart government actions they believe will damage the environment. Targets have included such agencies as the Bureau of Land Management and the U.S. Forest Service, as well as private companies and individuals. The supporters of Earth First! have sometimes been joined by the Animal Liberation Front and the Earth Liberation Front, along with small clusters of individuals acting independently.

Although there is an informal code of nonviolence, there is also a recognition that groups cannot be held responsible for what individuals decide to do, including violent confrontations and arson. Attacks on fur farms, restaurants serving meat, congressional offices, and ski resorts have blurred the line between radical causes and goals. Critics say the current generation of radicals read Abbey's book in school, become vegetarians, and get involved in protests. Then they become disillusioned and decide to take it a step further (Markels, 1999). Ecoterrorism, most agree, involves an unorganized, amorphous group who "join" by committing some act of sabotage or other covert action, and then christen themselves as radicals.

Initiating Litigation

In the early days of the environmental movement, firms devoted to natural resource litigation were relatively successful. For the most part, their cases were heard by judges who were sympathetic to their cause and to the concept of regulatory enforcement. That situation changed during the Reagan administration when new federal judges, appointed for life, brought a much more conservative view to the bench. These jurists believed in limited government involvement and were less interested in wilderness protection than resource exploitation. One president of a litigation group referred to the environmentalists' record as "one of profound disappointment"(Sale, 1993, 90), since much of their efforts was directed not at polluting industries but at federal agencies that had failed to implement and enforce the law. Traditional environmental lawyers became defensive, and business interests counterattacked (Sale, 1993; Dowie, 1995).

Litigation is still one of the most often used tools of activists. Lawsuits may take many forms, but the most common strategy is to sue an agency to force it to comply with the law. For instance, in 2001, four groups (Friends of the Earth, Environmental Working Group, Pesticide Action Network, and Pesticide Watch) won a favorable ruling in California against the state's Department of Pesticide Regulation (DPR). The groups alleged that the state agency had violated a 1989 law that required DPR to adopt clear and enforceable regulations for methyl bromide use by April of 1989. Methyl bromide is a highly volatile and acutely toxic biocide gas that is routinely used as a soil fumigant. It is an extremely dangerous compound that is known to cause nerve damage and birth defects in laboratory settings. One of the major concerns of the groups was that without state regulation, farmers could spray the chemical, and then spray and gas from the application could drift into surrounding neighborhoods. Although the court's ruling did not amount to a complete ban on the use of methyl bromide, the decision did require DPR to follow regulations, including a definite date by which new restrictions on use would be adopted.

Sometimes groups allege that the government itself is flouting the law. Under the 1992 Energy Policy Act, federal agencies were required to increase their purchase of alternative-fuel vehicles to 75 percent by the year 2000. The purpose of the statute is to reduce the nation's reliance upon oil, and it requires the federal government to serve as a model through its own vehicle purchases. Earthjustice, representing the Sierra Club, the Center for Biological Diversity, and the Bluewater Network, filed a lawsuit in 2002 against eighteen federal agencies that failed to comply. The agencies included the Department of the Interior, the Department of Commerce, and the Department of Energy, the agency responsible for enforcement of the law.

Even the threat of a long, protracted legal battle can be effective. Residents in the Jackson Hole, Wyoming, area hired well-known lawyer Gerry Spence to represent them when they learned that the U.S. Department of Energy was planning to build a nuclear waste incinerator just ninety miles away from them. The eastern Idaho facility had attracted little attention at first, but residents mounted a nationwide campaign that raised nearly $1 million for the group, Keep Yellowstone Nuclear Free. The legal wrangling over permits potentially could have tied the government up for years, so the department settled the suit instead, paying the group $150,000 in legal fees and putting the incineration project on indefinite hold.

Environmental opposition groups are not without their own resources, however. The Mountain States Legal Foundation, founded in 1977 to give corporate interests a legal voice, was initially funded by millionaire conservatives such as Joseph Coors. James Watt, who would become secretary of the interior under President Reagan, earned his environmental credentials working with the foundation. But generally, these litigants are less visible than individual corporations that are sued directly by the more ubiquitous environmental defense funds.

Co-optation and Greenwashing

Industry groups have attempted to influence environmental policy in ways that are not always available to traditional organizations. Through the strategy of co-optation, various corporations have served as sponsors for projects that might help them be perceived as environmentally friendly and aware. Firms such as the Shaklee Companies and Apple Computers gave money or equipment to help sponsor Earth Day activities in several cities, and organizers readily accepted their contributions, sometimes giving them credit as event sponsors and supporters. Others, such as Teva, which makes sandals and outdoor shoes, use their logo and contributions as a marketing tool to encourage consumers to buy their products.

Sometimes the process is not so subtle. In Texas, for example, Dow Chemical, Budweiser, and Copenhagen snuff serve as sponsors for various state agencies that are hungry for the cash such sponsorship brings to the state treasury. The state's Park and Wildlife Department adopted the Chevrolet Suburban as its "official SUV" when the company donated two vehicles and $230,000 to the agency. Despite the vehicles' poor reputation as being gas guzzlers with low fuel efficiency, the company now uses the department's logo in its ads.

Greenwashing is a term used to describe the public relations strategies that seek to put a more positive spin on companies that have caused environmental damage or on advertising that "paints" a business's activities as "green." Often, the advertising for environmental projects is totally separate from its product or service campaign. The efforts are designed to both restore or establish an environmental image. Some companies, such as General Electric, a supporter of the Audubon Society, have bought credibility by making contributions to major environmental organizations. Others have actually initiated projects that are ecologically sound practices. McDonald's, for example, changed its clamshell packaging to a less harmful material and was followed by other fast-food chains. In 1989, the Atlantic Richfield Company heralded its "environmental gasoline" that had been marketed as a replacement for more harmful leaded gas. Others adopted terminology such as *biodegradable* or *nontoxic* to appeal to consumers, at least until the Federal Trade Commission stepped in and began adopting standards for what had become dubious claims about safety and environmental protection.

Summary

This overview of environmental activism, providing a historical background that covers the past 100 years, includes information about the types of activist groups, their philosophies and their leadership, and many of the tactics and strategies that have been used to influence environmental policymaking. Although mainstream groups established in the first half of the twentieth century have dominated much of the political debate, newer, smaller, grassroots organizations and their members have played a significant role in topic- or area-specific issues. The more successful activists have combined a variety of strategies, depending upon their resources, both financial and expertise. Their efforts have sometimes been countered by groups representing the environmental opposition, which borrow the same tactics to work against the environmental groups. As activism becomes more professionalized and focuses on national and global issues, community groups and organizations have spent their energies on problems more directly related to their neighborhoods. Each type of group has shaped the environmental agenda and recruited members, some of whom overlap, form coalitions, and work in concert with one another.

Sources and Further Reading

Abbey, Edward. 1975. *The Monkey Wrench Gang.* New York: J. B. Lippincott.

Allen, Thomas B. 1987. *Guardian of the Wild: The Story of the National Wildlife Federation.* Bloomington: Indiana University Press.

Bailes, Kendall E. 1985. *Environmental History.* Denver, CO: University Press of America.

Bramwell, Anna. 1989. *Ecology in the 20th Century: A History.* New Haven: Yale University Press.

Brown, Michael H. 1987. *The Toxic Cloud: The Poisoning of America's Air.* New York: Harper and Row.

Bullard, Robert D. 2000. *Dumping in Dixie: Race, Class, and Environmental Quality.* Boulder, CO: Westview Press.

Caldwell, Lynton K. 1960. *Citizens and the Environment.* Bloomington: Indiana University Press.

Carrels, Peter. 1999. *Uphill against Water: The Great Dakota Water War.* Lincoln: University of Nebraska Press.

Carson, Rachel. 1962. *Silent Spring.* Boston: Houghton Mifflin.

Cawley, R. McGreggor. 1993. *Federal Land, Western Anger: The Sagebrush Rebellion and Environmental Politics.* Lawrence: University Press of Kansas.

Clarren, Rebecca. 2000. "A Bird? A Plane? It's the Environmental Air Force." *High Country News* (4 December): 6.

Cohen, Michael P. 1988. *The History of the Sierra Club, 1892–1970*. San Francisco: Sierra Club Books.

Cohen, Richard E. 1992. *Washington at Work: Back Rooms and Clean Air*. New York: Macmillan.

Commission for Racial Justice. 1987. *Toxic Waste and Race in the United States*. New York: Commission for Racial Justice.

Coolidge, Shelley Donald. 1996. "Boycotts of Companies Grow as a Protest Tool." *Christian Science Monitor* 88, no. 246 (15 November): 1.

Cooper, Mary H. 1995. "Environmental Movement at 25." *CQ Researcher* (31 March): 275–291.

Devall, Bill. 1988. *Simple in Means, Rich in Ends: Practicing Deep Ecology*. Salt Lake City: Peregrine Smith.

Devall, Bill, and George Sessions. 1985. *Deep Ecology*. Salt Lake City: Gibbs M. Smith.

Diamond, Irene, and Gloria Orenstein, eds. 1990. *Reweaving the World: The Emergence of Ecofeminism*. San Francisco: Sierra Club Books.

Dowie, Mark. 1995. *Losing Ground*. Cambridge: MIT Press.

Dunlap, Riley, and Angela G. Mertig, eds. 1992. *American Environmentalism: The U.S. Environmental Movement, 1970–1990*. Philadelphia: Taylor and Francis.

Durnil, Gordon K. 1995. *The Making of a Conservative Environmentalist*. Bloomington: Indiana University Press.

Easterbrook, Gregg. 1995. *A Moment on the Earth: The Coming Age of Environmental Optimism*. New York: Viking.

Echeverria, John, and Raymond Booth Ely. 1995. *Let the People Judge: Wise Use and the Private Property Rights Movement*. Covelo, CA: Island Press.

Floyd, Charles F., and Peter J. Shedd. 1979. *Highway Beautification: The Environmental Movement's Greatest Failure*. Boulder, CO: Westview Press.

Foreman, Christopher H., Jr. 1998. *The Promise and Peril of Environmental Justice*. Washington, DC: Brookings Institution Press.

Foreman, Dave. 1988. *Ecodefense*. Rev. ed. Tucson, AZ: Ned Ludd Books.

Fox, Stephen R. 1981. *John Muir and His Legacy: The American Conservation Movement*. Boston: Little, Brown.

Gibbs, Lois. 1982. *Love Canal: My Story*. Albany: State University of New York Press.

Glazer, Penina M., and Myron P. Glazer. 1998. *The Environmental Crusaders: Confronting Disaster and Mobilizing Community*. University Park: Pennsylvania State University Press.

Gottlieb, Robert. 1993. *Forcing the Spring: The Transformation of the American Environmental Movement.* Washington, DC: Island Press.

Gottlieb, Robert, and Helen Ingram. 1988. "The New Environmentalists." *The Progressive* 52: 14–15.

Graham, Frank, Jr. 1971. *Man' s Dominion: The Story of Conservation in America.* New York: M. Evans.

Grossman, Richard. 1988. *And on the Eighth Day We Bulldozed It.* San Francisco: Rainforest Action Network.

Hamilton, Cynthia. 1993. "Women, Homes, and Community: The Struggle in an Urban Environment." In Peter C. List, ed., *Radical Environmentalism: Philosophy and Tactics,* 221–227. Belmont, CA: Wadsworth Publishing.

Hays, Samuel P. 1959. *Conservation and the Gospel of Efficiency: The Progressive Conservation Movement, 1890–1920.* Cambridge: Harvard University Press.

———. 1987. *Beauty, Health, and Permanence: Environmental Politics in the United States, 1955–1988.* New York: Cambridge University Press.

Hynes, H. Patricia. 1989. *The Recurring Silent Spring.* New York: Pergamon Press.

Kerwin, Cornelius M. 1999. *Rulemaking: How Government Agencies Write Law and Make Policy.* Washington, DC: Congressional Quarterly Press.

Kline, Benjamin. 2000. *First along the River: A Brief History of the U.S. Environmental Movement.* San Francisco: Acada Books.

Kraft, Michael E. 2000. "U.S. Environmental Policy and Politics: From the 1960s to the 1990s." In Otis L. Graham Jr., ed., *Environmental Politics and Policy, 1960s–1990s,* 17–42. University Park: Pennsylvania State University Press.

League of Conservation Voters. 1996. *Biennial Report.* Washington, DC: League of Conservation Voters.

———. 2000. *National Environmental Scorecard.* Washington, DC: League of Conservation Voters.

Levine, Adeline. 1982. *Love Canal: Science, Politics, and People.* Lexington, MA: Lexington Books.

Manes, Christopher. 1990. *Green Rage.* Boston: Little, Brown.

Markels, Alex. 1999. "Backfire." *Mother Jones* (March–April): 60–64, 78–79.

McCormick, John. 1989. *Reclaiming Paradise: The Global Environmental Movement.* Bloomington: Indiana University Press.

Melosi, Martin V. 1981. *Garbage in the Cities: Refuse, Reform, and the Environment, 1880–1980.* College Station: Texas A & M University Press.

———. 2000. "Environmental Justice, Political Agenda Setting, and the Myths of History." In Otis L. Graham Jr., ed., *Environmental Politics and Policy, 1960s–1990s*, 43–71. University Park: Pennsylvania State University Press.

Merchant, Carolyn. 1980. *The Death of Nature: Women, Ecology, and the Scientific Revolution*. San Francisco: Harper and Row.

Milbrath, Lester W. 1984. *Environmentalists: Vanguard for a New Society*. Albany: State University of New York Press.

Mitchell, Robert Cameron. 1989. "From Conservation to Environmental Movement: The Development of the Modern Environmental Lobbies." In Michael J. Lacey, ed., *Government and Environmental Politics*, 82–113. Washington, DC: Wilson Center Press.

Mitchell, Robert Cameron, Angela G. Mertig, and Riley E. Dunlap. 1992. "Twenty Years of Environmental Mobilization: Trends among National Environmental Organizations." In Riley E. Dunlap and Angela G. Mertig, eds., *American Environmentalism*, 11–26. Philadelphia: Taylor and Francis.

Mohai, Paul. 1985. "Public Concern and Elite Involvement in Environmental Conservation." *Social Science Quarterly* 66: 820–838.

Nash, Roderick. 1968. *The American Environment: Readings in the History of Conservation*. London: Addison-Wesley.

Natapoff, Sasha. 1989. *Stormy Weather: The Promise of the U.S. Environmental Movement*. Washington, DC: Institute for Policy Studies.

"New Initiatives." *Boycott Action News*. http://www.cooperamerica.org/boycotts. Accessed 23 November 2001.

Oelschlaeger, Max. 1994. *After Earth Day: Continuing the Conservation Effort*. New Haven: Yale University Press.

Opie, John. 1998. *Nature's Nation: An Environmental History of the United States*. New York: Holt, Rinehart, and Winston.

Paehlke, Robert. 1989. *Environmentalism and the Future of Progressive Politics*. New Haven: Yale University Press.

Petulla, Joseph M. 1980. *American Environmentalism: Values, Tactics, Priorities*. College Station: Texas A & M University Press.

Rensenbrink, John. 1999. *Against All Odds: The Green Transformation of American Politics*. Raymond, ME: Leopold Press.

Rudig, Wolfgang. 1991. "Green Party Politics around the World." *Environment* 33, no. 8: 6–9, 29–31.

Sale, Kirkpatrick. 1993. *The Green Revolution: The American Environmental Movement 1962–1992*. New York: Hill and Wang.

Scarce, Rik. 1990. *Eco-Warriors: Understanding the Radical Environmental Movement*. Chicago: Noble Press.

Scheffer, Victor B. 1991. *The Shaping of Environmentalism in America.* Seattle: University of Washington Press.

Schwab, Jim. 1994. *Deeper Shades of Green: The Rise of Blue-Collar and Minority Environmentalism in America.* San Francisco: Sierra Club Books.

Shabecoff, Philip. 1993. *A Fierce Green Fire: The American Environmental Movement.* New York: Hill and Wang.

Shapiro, Martin. 1986. "APA: Past, Present, and Future." *Virginia Law Review* 72: 452.

Smith, Jeffery. 1997. "Evangelical Christians Preach a Green Gospel." *High Country News* 29, no. 8 (28 April). http://www.hcn.org/ servlets/hcn.URLRemapper/1997/apr28. Accessed 1 September 2000.

Smyth, Julie Carr. 2001. "Environmental Groups Protest Slow Pollution Enforcement." *The Plain Dealer* (14 June).

Strong, Douglas H. 1971. *The Conservationists.* Menlo Park, CA: Addison-Wesley.

Switzer, Jacqueline. 1997. *Green Backlash: The History and Politics of Environmental Opposition in the U.S.* Boulder, CO: Lynne Rienner.

Szasa, Andrew. 1994. *EcoPopulism: Toxic Waste and the Movement for Environmental Justice.* Minneapolis: University of Minnesota Press.

Talbot, Allan R. 1972. *Power along the Hudson: The Storm King Case and the Birth of Environmentalism.* New York: Dutton.

Tober, James A. 1989. *Wildlife and the Public Interest: Nonprofit Organizations and Federal Wildlife Policy.* New York: Praeger.

Tucker, William. 1982. *Progress and Privilege: America in the Age of Environmentalism.* New York: Doubleday.

Warren, Karen. 1993. "The Power and the Promise of Ecological Feminism." In Michael E. Zimmerman, ed., *Environmental Philosophy: From Animal Rights to Radical Ecology,* 320–341. Englewood Cliffs, NJ: Prentice-Hall.

Wolf, Hazel. 1993–1994. "The Founding Mothers of Environmentalism." *Earth Island Journal* (Winter): 36–37.

Worster, Donald. 1977. *Nature's Economy: The Roots of Ecology.* San Francisco: Sierra Club Books.

———. 1993. *The Wealth of Nature: Environmental History and the Ecological Imagination.* New York: Oxford University Press.

Young, John. 1990. *Sustaining the Earth: The Story of the Environmental Movement.* Cambridge: Harvard University Press.

2

Problems, Controversies, and Solutions

The ways in which environmental activists get involved are as numerous as there are environmental problems to solve. For example, although it is true that no one is "for" or supportive of air pollution, there are controversies as to how best to deal with air quality issues. Some groups have sued the Environmental Protection Agency to force it to comply with and enforce the provisions of the 1990 Clean Air Act Amendments. Others believe the focus should be on controlling pollution from automobiles, and still others dedicate themselves to reducing the emissions from factory smokestacks. There are groups that want each state to be responsible for cleaning up its own air, and still others that see air pollution as a global problem.

This chapter explores some of those issues by examining the problems around which environmental activists coalesce and organize their interests. A statement of each problem and how it is perceived by various groups is followed by an explanation of the arguments and strategies activists have used. The objective is to provide an overall understanding of key controversies and how a variety of interests has approached them.

Battling for Water in the Klamath Basin

Although battles over water in the arid West have endured for more than a century, one of the more recent disputes illustrates what happens when competing sets of interests are pitted against

41

one another. One of those interests belongs to the members of the Klamath Tribes, who are trying to regain their culture, home-lands, and a primary source of food—native fish. A second interest is that of local farmers who claim water rights for agricultural use granted through the U.S. Reclamation Project trust agreement. They have protested the cutoff of irrigation water during a time of severe drought. About 250 miles downriver, where the Klamath River empties into the Pacific Ocean, irrigation diversions, cattle grazing, logging, and mining have led to staggering economic losses for fishing-dependent communities. And last, the U.S. Fish and Wildlife Service and the National Marine Fisheries Service have used their power under the Endangered Species Act to prioritize water flows into two major national wildlife refuges that are home to waterfowl and to waterways that previously were home to three species of fish.

The Klamath Basin, which spreads over parts of southern Oregon and northern California, has been the subject of controversy for nearly 140 years. In 1854, the three Klamath Tribes signed an agreement with the U.S. government that allowed them sufficient water to meet the tribes' needs in hunting, gathering, and fishing. The treaty has survived numerous court challenges, even after the federal government ended formal recognition of the tribes in 1954. After the government "terminated" the tribes, their land of nearly a million acres came under control of the Department of Agriculture and was opened to commercial logging. In 1986, the tribes regained federal recognition and began the process of trying to reclaim their lands. The populations of fish they had traditionally eaten for sustenance had decreased over the years, and by 1986, the tribes stopped fishing for them. Two years later, the number of fish was so low that two species of endangered sucker fish and the threatened coho salmon could not be sustained.

Just after the turn of the twentieth century, the U.S. Bureau of Reclamation approached the Klamath Basin as a giant plumbing project: rerouting rivers, building dams, and draining wetlands so that homesteaders would be convinced to farm about 200,000 acres the government considered to be worthless swamps. The area is actually high desert, however, and too alkaline for agriculture, reliant upon an unreliable snowpack as a water source. By changing the region's hydrology, the government did not consider the significance of its actions on local plants and animals, nor was there any attempt to meet the agreements that had been reached with the Klamath Tribes.

The problem this presents is best expressed by Allen Foreman, chairman of the 3,300-member Klamath Tribes of Oregon, in speaking at a hearing before the Senate Energy and Natural Resources Committee on 21 March 2001: "The federal and state governments have overtaxed water supply by promising too much water to too many people" (Foreman, 2001).

The dispute, and others like it, has involved several forms of political activism. In 1998, for example, twenty-nine irrigation districts in the Central Valley of California near Klamath sued the federal government, claiming that it unfairly took $25 million worth of water. A federal judge ruled in May 2001 that the federal government is prohibited from taking private property (the water) without paying the farmers for it. One rancher noted, "By taking our water they made our land useless" (Clarren, 2001). The California group hopes to win as much as $1 billion from the federal government.

Initially, farmers in Oregon did not have the time or resources to become involved in lengthy litigation. The federal government turned off the spigot for 200,000 acres of farmland in April 2001. Farms suffered not only from the lack of irrigation but from a severe drought that brought despair to small, family-owned operations that raise potatoes, alfalfa, and cattle. The region received less than a quarter of the usual mount of rainfall in winter 2000, and even an emergency delivery of water could not help. The federal government released about 76,000 acre feet of water into the basin for about a month during the summer but cut off the supply in late August 2001. A subsequent attempt to run an irrigation pipe from the Upper Klamath Lake to a small spillway was actually more symbolic than it was valuable, since the amount of water that can be pumped from the lake is insufficient to save any of the farmers' crops.

During the dispute, over 15,000 farmers and their supporters held a protest rally in Klamath Falls, Oregon. Others set up a camp alongside a trailer command post at the irrigation headgates, participating in their own rural form of civil disobedience. On at least four occasions, the group pried open the irrigation system's headgates to allow water to flow freely, so U.S. marshals were brought in to close the gates. In support of the farmers, a Convoy of Tears that included trucks and cars from Nevada, Montana, and southern California headed toward Klamath Falls, bringing canned food, livestock, and monetary donations.

In order to capture media attention and get even more public support, farmers held a bucket brigade down the main street of Klamath Falls, with some supporters bringing in buckets of their own water from out of state. The protest included a group of riders on horseback, waving American flags; others framed the controversy as one over private property rights. Another group of 300 protesters, members of Farmers Against Regulatory Madness (FARM), climbed a security fence that had been erected at the headgates so they could personally deliver ownership documents to officials from the Bureau of Land Management. A day after the September 11, 2001, terrorist attack on the United States, they withdrew and called a truce to the protest, telling officials they did not want to cause more problems for the federal government.

The strategies employed by these varying interests appear to have had an impact on federal officials, who had previously sought to have two lawsuits filed over the water allocation settled through mediation. Interior Secretary Gale Norton asked the National Academy of Sciences to review the opinions by biologists from the U.S. Fish and Wildlife Service that were the foundation for the decision to cut off the irrigation water to the farmers. On 11 September 2001, the Interior Department allowed 6,300 acre feet of water to flow to the lower Klamath National Wildlife Refuge, so that migrating birds would have a place to rest. Six months later, Interior Secretary Norton and Agriculture Secretary Ann Veneman opened the headgates into an agricultural canal after a decision was reached to authorize the release of two months of irrigation water. Winter snow and rain had slightly increased water supplies, although Indian activists carried a banner that read "Bush kills salmon." Meanwhile, the American Land Conservancy offered a plan to buy the land and water rights in the Klamath Basin, which would include a site for storing 100,000 acre feet of water to balance the needs of agriculture with the needs of fish and wildlife.

In May 2002, a federal judge ruled against the Pacific Coast Federation of Fishermen's Associations, a group that had filed suit against the U.S. Bureau of Reclamation and the National Marine Fisheries Services in April 2002. The federation's attorneys argued in U.S. District Court that the water that had been stored for use by the farmers should be released in order to help young salmon migrate to the ocean. In their suit, the plaintiffs said that the river's flows were as little as 60 percent of the water needed by salmon to survive. U.S. Forest Service employees had been trying to rescue hundreds of salmon and other fish who were stranded in puddles along the banks of the Klamath. But the judge said that there was insufficient evidence that the salmon needed more water, ruling in favor of the farmers.

Some farmers believe the court and federal agencies' actions were too little, too late. For the members of the Klamath Tribes, there is little indication that their interests will be taken into consideration in the twenty-first century any more than they were in the nineteenth. For the federal agencies involved in the water diversion, there is additional skepticism and scorn over what some believe are bureaucratic decisions made at the expense of individuals. For the endangered species, waterfowl, migratory birds, fish, and other wildlife, the battle for water is not yet over.

Sources and Further Reading

Boxall, Bettina. "Tribes Say Injustice Flows with Water for Farms." *Los Angeles Times*, 10 September 2001.

Clarren, Rebecca. "Will Farmers Harvest a Legal Take?" *High Country News*, 13 August 2001.

Foreman, Allen. Testimony before the Senate Committee on Energy and Natural Resources, Subcommittee on Water and Power, Oversight Hearing on the Klamath Reclamation Project, 21 March 2001.

Kemper, Steve. "A Flap over Water." *Smithsonian* 32, no. 6 (September 2001): 20–24.

Knickerbocker, Brad. "Farmers and Ranchers Unite." *Christian Science Monitor*, 27 August 2001.

Wilson, Robert. "Klamath's Federal Agencies Map Different Realities." *High Country News*, 13 August 2001.

Zuckerman, Seth. "Klamath Water Wars: Systems out of Sync." *Christian Science Monitor*, 27 July 2001.

Buying Conservation in the Southwest

There's an old saying, "if you can't beat 'em, join 'em," that can be applied to a contentious issue in the American Southwest. In Arizona, the state supreme court ruled in 2001 that the Land Department could not refuse to issue grazing leases to environmental groups who planned to "rest the land" rather than grazing animals. The Land Department, created as a trustee for land given to Arizona by the federal government when it became a state, is required to pick the "highest and best" bidder for grazing leases on state lands. The idea was to maximize income for the state and protect the value of the land.

But a New Mexico–based group, Forest Guardians, decided to use the system in a slightly different way by placing its own bid and then restoring the land, an action that one land commissioner called "a risky precedent" (Davenport, 2001). He accused the state court of effectively neutering legislative procedures that are in place to keep the commission honest in appraising a property's value and classification. An attorney for the group responded that there were two benefits to the lease under Forest Guardians: overgrazed land could be restored, and the state's trust beneficiaries, such as schools, would end up getting additional income through acceptance of a higher bid. Forest Guardians was joined in the litigation in a brief filed by the 30,000-member Arizona Education Association. Its director called for leasing agreements that resulted in the highest income for the school trust fund.

Opponents to the ruling, such as the Arizona Cattlemen's Association, argued that some ranchers might not bid on grazing leases in the future. Because the lease agreements require that fences and other infrastructure be maintained, the concern was that groups who used the land for nongrazing purposes would not take care of what was already in place. In his dissenting opinion, one justice called the ruling "bad law" that might result in unintended consequences.

Supporters of this type of activism use the term *conservation leasing* to describe what Forest Guardians is attempting to do. Rather than allowing ranching interests, many of whom have held grazing leases for decades, to control land use, they believe that market-based solutions can serve as a mechanism for ecological restoration of state lands. Groups such as Environmental Defense note that the strategy can also be used by private landowners, who can increase the income and value of their land by agreeing not to develop it in order to protect endangered species or habitat. Similar strategies include safe harbor policies and agreements, conservation banking, and other positive incentives for landowners, rather than relying upon goodwill and voluntary efforts.

In New Mexico, for example, there are an estimated 9.2 million acres of state trust land that, like Arizona, are to be used as an income source for such institutions as schools, universities, and hospitals (State Lands Conservation Leasing, 2001). Forest Guardians prepared a biological survey of the state trust lands as a way of developing a plan for restoration and diversity protection. Their goal is to lease as many as 100 parcels currently being grazed, including those where there are imperiled native grasslands. Its map of ecologically valuable sites serves as the basis for a long-term acquisition plan.

In these two examples, there has been a considerable amount of confrontation over the long-term institution of livestock grazing leases. But in other instances, collaborative land use planning has opened up a dialogue among farmers, ranchers, government officials, environmental groups, and local residents over how land should be used. The Northern New Mexico Collaborative Stewardship, organized by U.S. Forest Service officials in 1991, has brought together community interests, timber industry officials, Native Americans, and environmental advocates to survey local areas, taking into consideration economic needs, ecological dangers, and other factors. In Oregon, the Trout Creek Mountain Working Group, a private collaborative group, organized and convinced ranchers to stop grazing for three years in areas where livestock had overgrazed land adjacent to riparian areas that are the habitat of the cutthroat trout. Other efforts throughout the West have made the terms *common ground* and *collaborative environmentalism* more than just wishful thinking.

Sources and Further Reading

Bean, Michael. "Endangered Species, Endangered Act?" *Environment* 41, no. 1 (1999): 13–18, 34–38.

Davenport, Paul. "Supreme Court Rules Conservationists Can Lease Grazing Land for 'Resting.'" *Arizona Capitol Times*, 30 November 2001, 26.

Forest Guardians. "Unranching the Southwest." http://www.fguardians.org/consleas.html. Accessed 23 December 2001.

Knudson, Tim. "The Ranch Restored: An Overworked Land Comes Back to Life." *High Country News* (1 March 1999): 13–16.

State Lands Conservation Leasing. http://www.fguardians.org/consleas.html. Accessed 23 December 2001.

Williams, Ted. "The New Guardians." *Audubon* 101, no. 1 (1999): 34–39.

Can the Green Party Survive in the United States?

Green parties developed and continue to develop because of a desire and need to view the world differently than popular political thought. It is difficult to sum up all that the Greens, as they are known, stand for, but generally they organize around ideas of social justice, peace, and an antinuclear agenda. Carrying these values with them, green political parties are an important addition to a democratic system, although there are questions over whether or not they will survive.

The evolution of the green party movement begins in Tasmania in March 1972. The world's first green political party was known as the United Tasmania Group, but they later changed their name to the Tasmanian Greens. The group has worked with government officials to support firearms control, mining law amendments, natural death legislation, and the formation of a nuclear-free Tasmania. The Values Party, now the Greens of New Zealand, was also established in 1972, focusing on environmentally sustainable economic development and grassroots campaigns.

In Europe, the first green party was founded in the United Kingdom (UK) in 1973. Originally called People, and then the Ecology Party, and finally the Green Party, the UK Greens were followed quickly by similar groups in Switzerland and Germany. The first green party member elected to a national parliament was a Swiss politician (1979), and in 1983, Germany's Die Grünen passed the 5 percent threshold of votes that allowed it to gain seats in the Bundestag.

Internationally, there are green parties on virtually every continent, although primarily in countries with a representative form of government. They share ten "key values" that form their philosophical base: ecological wisdom, social justice, grassroots democracy, nonviolence, decentralization, community-based economics, feminism, respect for diversity, personal and global responsibility, and sustainability.

In the United States, the Green Party continues to struggle to survive. Influenced by the European Greens, the first organizational meetings were held in 1984, and by 1992, several state-level green parties had gained ballot access. The first national nominating convention was held in Los Angeles in 1996, with Ralph Nader and Winona LaDuke as presidential and vice presidential nominees. Nationwide, the party gained about 700,000 votes—1 percent of the vote (Rensenbrink, 1999). In 2000, the two candidates were nominated again and were on the ballot in forty-five states, more than twice the number as 1996. This time, the Green Party won 2.7 million votes, or about 3 percent of those cast

(Pollock, 2001). But the closeness of the election, and the failure of Democratic candidate Al Gore, who had the support of the majority of environment groups, cast a shadow on the future of the Greens.

Most of their successes have been at the local level. In 1996, in Arcata, California, the Green Party won a majority of seats on the city council, but the victory lasted only two years. In 2000, Sebastopol, California, became the second U.S. city to form a Green Party majority within a city council, with one member appointed as mayor. Only one Green Party candidate, Audie Bock, has been elected to the state legislature, serving a short term in the California Assembly. Of the fifty-seven Green Party candidates running for Congress in 2000, none were successful.

There are several obstacles that do not bode well for the U.S. Greens. In order to gain national party status, the Green Party must nominate candidates for various federal offices in numerous states, engage in activities such as voter registration drives on an ongoing basis, and hold a national convention. In both 1996 and 2000, the Greens were unable to meet that criteria, limiting the amount of contributions the party could accept. In addition, in the U.S. system, most elections are decided on the basis of which candidate gets the plurality of the vote. So even in races where there is electoral competition among several candidates, the lack of proportional representation shuts out even those who come close to winning. Still, the Green Party claims it is the fastest growing political party in the United States, with over 140,000 members in California alone.

Sources and Further Reading

Booth, William. "Green Party's Radically New Role—Governing." *Washington Post,* 14 March 1997.

Brooke, James. "Green Party Grows, So Does Democrats' Dismay." *New York Times* (30 July 1998): A12.

Pollock, Danny. "Green Party Seeks National Status." 31 July 2001. http://www.greens.org/california/news. Accessed 20 September 2001.

Poole, William. "Inside Ecotopia." *Sierra* (January–February 1998): 30–33.

Rensenbrink, John. *Against All Odds: The Green Transformation of American Politics.* Raymond, ME: Leopold Press, 1999.

Does Ecoterrorism Work?

At first, most of the tactics involved nonviolent direct action, such as a sit-in on a roadway to stop logging trucks from entering publicly owned land. Then the strategies began to move toward vandalism. Logging company vehicles had dirt poured into their engines. Slogans were spray-painted on the walls of fast-food restaurants. Locks on government buildings were glued shut. By the late 1980s, several groups of environmental activists escalated their forms of protest to burglary, arson, threats of violence, acts of major destruction, and perhaps even murder. U.S. Forest Service and Bureau of Land Management workers have testified that they suffer from anxiety and fear for their personal safety as the violence has spread.

This form of activism goes by several names, such as monkey wrenching (based on the title of a book by Edward Abbey), ecotage, ecodefense, or more commonly, ecoterrorism. Supporters are accused of using siege warfare that saps local law enforcement resources and ties up the legal system, depriving citizens of the essential functions of government. Are these tactics successful in affecting policy change? This controversy has plagued organizations along the full spectrum of the environmental movement for more than two decades and remains unresolved.

Ecoterrorism is often associated with the loosely knit protest group, Earth First!, which was formed in 1979. Its supporters believed that activism by mainstream organizations was not moving along fast enough and that more radical strategies were called for. Other groups such as the Earth Liberation Front and the Animal Liberation Front have also supported a more militant action agenda, and the latter two groups often work cooperatively.

In a 1998 congressional hearing on crime, subcommittee chairman Bill McCollum of Florida noted that the message of environmental protesters gets lost when injury or death result. He said that ecoterrorism by radical and violent groups only encourages fear and anger, even when it is conducted to protect the earth. One of those directly affected was Frank Riggs, who represents parts of northern California in the House of Representatives. On 16 October 1997, his district office in Eureka was assaulted by North Coast Earth First! activists, several of whom were dressed completely in black and wore hooded ski masks or dark goggles to hide their identities. Riggs said the intruders began by dumping a 500-pound tree stump onto his office floor and then dumped garbage bags filled with sawdust and pine needles all over his office. He said that his employees were traumatized by the incident, and one appeared before the congressional subcommittee to describe her experience. Later, Earth First! members gave media interviews claiming credit for the action at Riggs's office. They said they were protesting the government's efforts to acquire the privately owned Headwaters Forest in Humboldt County, California.

Other ecoterrorist acts are designed to garner media attention or embarrass government officials in a public setting. The masked individuals who raided Congressman Riggs's office videotaped the entire incident. In 1997, a member of Earth First! threw bison entrails on Senator Conrad Burns, Secretary of Agriculture Dan Glickman, and Montana governor Mark Rocicot. Like those in other types of terrorist groups, the activists rely upon unexpected hit-and-run attacks.

Pima County Estates, a 471-acre residential development, borders the Coronado National Forest in southern Arizona near Tucson. The 5,000- to 8,000-square-foot houses in the project back up to the Catalina Foothills, or at least some of them used to. In 1989, county officials decided to try to purchase 107 acres of open space in the canyon area and offered the developer $4.3 million for the land. But the parties could never agree on a purchase price, and eventually, multimillion-dollar homes began being built, mostly on 1- to 5-acre lots. Four vacant homes were torched by an arsonist on 12 June 2001. Although a local Earth First! spokesperson referred to the fires as "a positive thing" (Davis and Tobin, 2001), no group claimed responsibility.

To counter these attacks, both private individuals and public officials have taken a number of steps to protect themselves. The president of Alliance for America, who owns a small family logging operation in Montana, began wearing a bulletproof vest, and his children's school worked out a system to move them to safety when threats were made. Many other targets have been granted additional law enforcement security, although some who have asked for protection have been ignored. Some law enforcement officials have said they are intimidated by the violence and threats and fear reprisals.

Conservative groups and some members of Congress have sought to curb ecoterrorism through the use of existing mechanisms, such as the Racketeer Influenced and Corrupt Organizations (RICO) statute. Some believe that the scope of the law could be expanded to cover these types of actions, even though it was initially enacted to deal with organized crime. Others have sought state bans on the locking devices protesters use to handcuff themselves inside steel pipes to resist their arrest. One group favored an amendment to the Resource Enterprise Protection Amendment of 1998 so that loggers, miners, farmers, food processors, and others would be covered. The original law, the Animal Enterprise Protection Act of 1993, federalized crimes or property damage over $10,000 or that resulted in dismemberment or death to a human being as a result of attacks on animal enterprises. Other critics of the law believe it is not being fully enforced and is virtually ineffectual.

But there is another side to the controversy. Although there is no support of activism that results in injury or death, mainstream environmental organizations believe that the militant groups serve as an important foil for their efforts. By pursuing a radical or militant strategy, they distance themselves from more traditional groups such as the Wilderness Society or the Izaak Walton League. The policy agendas of those groups start to look moderate in comparison.

Sources and Further Reading

Best, Allen. "Vail Fires Outrage Community." *High Country News*, 9 November 1998.

Davis, Tony, and Mitch Tobin. "Luxury Homes Torched in Tucson." *High Country News* (2 July 2001): 3.

Denson, Bryan, and James Long. "Eco-Terrorism Sweeps the American West." *The Oregonian*, 26 September 1999.

Foreman, Dave. *Confessions of an Eco-Warrior.* New York: Harmony Books, 1991.

Sunde, Scott, and Paul Shukovsky. "Elusive Radicals Escalate Tactics in Nature's Name." *Seattle Post-Intelligencer*, 18 June 2001.

U.S. House. Subcommittee on Crime of the Committee on the Judiciary. *Hearing: Acts of Ecoterrorism by Radical Environmental Organizations.* 105th Cong., 2d sess., 9 June 1998.

Hogs and Health in North Carolina

Ever wonder where the bacon in your bacon-and-eggs breakfast comes from? Chances are, it came from a factory pork production plant in North Carolina. There are more than 2,600 registered hog operations in the state, raising an estimated 10 million hogs. Pork production also involves the creation of open-air pits called "lagoons" where hog wastes are stored, with most of the 3,800 pits dug near waterways or adjacent lands that drain into coastal waters. Over 500 of the pits are deemed abandoned, and the problem is especially sensitive in eastern North Carolina, where hog farming is the most intense and concentrated (Environmental Defense Fund, 1999).

There is an environmental price to be paid for this type of farm production. Nineteen million tons of manure are produced each year, resulting in threats of contamination to groundwater and drinking water, emissions of atmospheric ammonia nitrogen and accompanying foul odors, and a reduced quality of life for residents and visitors who add an estimated $2 billion to the state's economy from coastal tourism (Environmental Defense Fund, 1999).

Additional concerns have been raised over the treatment process itself, which may allow disease-causing bacteria to be released—an important environmental health hazard. Studies have linked hog factories to increases in respiratory illnesses such as asthma and to levels of fecal coliform that exceed both state and federal standards.

The state's lagoons must be built to comply with standards developed by the Natural Resources Conservation Service, although regulations enacted in 1993 have been shown to be insufficient to prevent groundwater contamination. Leakage from an average-size three-acre lagoon could amount to as much as 1 million gallons a year. Airborne emissions from hog farms are not regulated by the state or by the federal Environmental Protection Agency. Despite rules that require owners to manage lagoons even after they have been abandoned, the state's environmental agencies admit that many are left unattended, and there are no regulations that indicate when an inactive site must be closed.

The issue has brought together a variety of environmental activists and organizations, as well as strong opposition from the hog farm industry. The North Carolina Public Interest Group, a state-level representative of the national Public Interest Research Group (PIRG), started its Campaign to Clean Up Corporate Hog Farms in the late 1980s with a set of recommendations to the governor of North Carolina and the state's General Assembly. The group called for a plan to phase out the lagoons and a ban on the rebuilding and restocking of operations after a natural disaster, such as Hurricane Floyd. These efforts were underscored by a study conducted by Environmental Defense's North Carolina office that produced a report calling for a "solutions package" that includes the development of performance standards for existing farms and the strengthening of standards for closing lagoons. Environmental Defense also called upon the state to look for alternative waste systems and new technology that would reduce harmful environmental impacts. The study relied upon researchers at North Carolina State University, the University of North Carolina, and Kansas State University and on the state's own data to make its case.

Several other mainstream strategies were used to put the issue on the political agenda. Public education on the hog lagoon problem has been ongoing. Environmental Defense created a web site called Hog Watch so that the situation could be publicized and updated, as well as being accessible to activists around the country. Two lawsuits were filed against Smithfield Foods, Inc., the largest hog producer and pork processor in the world, uniting the Southern Environmental Law Center, the North Carolina chapter of the Sierra Club, and Environmental Defense. Smithfield responded by issuing press releases that characterized the plaintiffs as "contingency-fee lawyers" who were using a "campaign of extortion-by-litigation" that "represent a gross perversion of our legal system" ("Smithfield Foods Reacts to Lawsuits, 2001"). The company singled out Robert F. Kennedy Jr. and cited the "social agenda of a group of unelected, unaccountable trial lawyers who want to override the legal and regulatory framework, substitute their own personal desires and financial interests for the wishes of the people and enrich themselves at the hands of the pork industry and ultimately the American consumer" ("Smithfield Foods Reacts to Lawsuits, 2001"). The industry also used its resources to spend $1.1 million in lobbying, ads, and campaign contributions in North Carolina, according to the research group Democracy South. The industry accused "polluter front group" ("Protect Our Waters, 2001") Farmers for Fairness of spending $2.6 million in 1998 to fund three-fourths of the campaign to unseat a pro-environmental legislator.

The rhetoric and litigation had two results. In July 2000, North Carolina's attorney general reached an agreement with Smithfield Foods that included $15 million for North Carolina State University to research alternative waste treatment. Once a successful alternative is agreed upon, which must be within two years after the commencement of the study, the company must convert its operations within three years and must eliminate all of its hog lagoons within five years. Litigation against the company was a less successful strategy. In March 2001, a Superior Court judge in Raleigh, North Carolina, ruled that the Waterkeepers Alliance, led by Kennedy, lacked standing to sue Smithfield and dismissed the two lawsuits because landowners had failed to state a single claim.

In one sense, environmental activists might claim credit for the Smithfield agreement, noting that by creating a coalition and backing up their claims with scientific research, they forced the state to take action after years of ignoring the problem. Smithfield might also consider itself a winner because it avoided litigation and portrayed Kennedy and his supporters as naive "unelected, unaccountable trial lawyers" who "declared war on an industry that constitutes a way of life across the country" ("Smithfield Foods Reacts to Lawsuits, 2001").

Sources and Further Reading

Environmental Defense. http://www.hogwatch.org.

Environmental Defense Fund. *Hog Lagoons: Pitting Pork Waste against Public Health and Environment.* Raleigh, NC: Environmental Defense Fund, 1999.

"Environmentalists Applaud Action Requiring Smithfield Foods to Eliminate NC Hog Lagoons" (25 July 2000 news release). http://www.selcga.org/archive.shtml. Accessed 29 December 2001.

"Protect Our Waters: NCPIRG's Campaign to Clean Up Corporate Hog Farms." http://www.ncpirg.org/hogs. Accessed 29 December 2001.

"Smithfield Foods Reacts to Lawsuits" (28 February 2001), and "Court Throws Out Environmental Suits against Smithfield" (29 March 2001 news release). http://www.smithfieldfoods.com/news/archives.html. Accessed 29 December 2001.

Justice in Warren County

In 1978, a North Carolina firm, Ward Transformer Company, hired New York liquid trucking company owner Robert Burns and his two sons to dispose of 31,000 gallons of contaminated oil from its operations. At the time, the Environmental Protection Agency (EPA) had restricted the sale of waste that was contaminated with polychlorinated biphenyls (PCBs), which were classified as carcinogens, or cancer-causing substances. It is known that the owners of the trucking firm and Ward Transformer Company decided to illegally discharge the liquid waste at night as the truck drove along the dirt shoulders of more than 200 miles of roadways. Citizens began to complain about the foul odors and minor health complications, and a subsequent investigation by the state found the transformer company guilty and arrested the three men for illegal dumping. North Carolina officials filed a $12.5 million lawsuit against both firms.

This did not end the story, however. The state realized that the PCB-contaminated dirt from the roads had to be dealt with in some way. They proposed to dump the material at a location in the small rural community of Afton in Warren County, even though the land did not meet the EPA's operating requirements for a hazardous waste landfill. The state purchased the property from a farmer who was facing foreclosure and bankruptcy, convinced that the landfill could be designed to hold the contamination indefinitely.

In 1980, Warren County's population was 65 percent black, and it was the 97th poorest county in the state (out of 100 counties) with a per capita income of $5,000 (McGurty, 2000). The local residents found out about the proposed dumping and filed two lawsuits to stop the project, one with the legal and financial assistance of the National Association for the Advancement of Colored People (NAACP). Both suits were rejected in federal court.

Changing tactics, a group of primarily white landowners sought the help of a black Baptist pastor who assisted in contacting local civil rights leaders to form a grassroots group to oppose the dumping, Warren County Citizens Concerned About PCBs. They advocated a safer site, recommending an existing successful chemical waste facility in Emelle, Alabama, one of three legal national dumping sites in the nation. Ironically, the Alabama facility, the largest hazardous waste site in the United States, was also in a predominantly black and poor area.

The coalition between whites and blacks gave the group access to a larger base of support and legitimized the campaign. Locally, two churches helped to disseminate information about activities to residents, and the church leaders called upon their brethren in other church networks for support. The situation began to attract the attention of several other organizations, including the United Church of Christ Commission for Racial

Justice, the Black Congressional Caucus, the Southern Christian Leadership Conference, the Congress of Racial Equality, and civil rights leaders in other states. They believed that the Warren County site had been selected because it was a poor, minority community rather than the best location from an environmental perspective. The local chapter of the NAACP argued that because the majority of the residents were poor, they were politically powerless. The actions of the state were said to be racially motivated and became exemplary of what would become known as environmental or toxic racism.

When the state began the cleanup operation and started hauling the 6,000 truckloads of contaminated dirt to the landfill, residents and their supporters began a four-week period of civil disobedience from 15 September to 12 October 1982 by placing their bodies in the roadway where the trucks hauling the material would have to travel. Using the traditional tactics of the civil rights movement, protesters disrupted construction, and over 500 persons were arrested. Prayer became a regular part of any protest, bringing white and black citizens together. The groups also organized a sixty-mile march from Warren County to the state capitol in Raleigh, where they presented their cause in public. Their protests were unsuccessful, and the project was eventually completed. Realizing that the citizens of Warren County were still outraged, the governor, James Hunt, wrote a letter to residents. He promised that through a joint local-state-federal working group, the state would push as hard as it could for the detoxification of the landfill when and if the appropriate and feasible technology is developed. That promise was made harder to fulfill in 1992, when it was discovered that 1.5 million gallons of contaminated water were trapped inside the landfill. An estimated $25 million has been spent by the state of North Carolina to try to clean up and detoxify the Warren County landfill.

Throughout the 1990s, Warren County's citizens' groups continued to call for more investigations and studies of the siting and operation of the landfill. Documents showed that from a scientific standpoint, the Warren County facility was not the most suitable place to dump the contaminated dirt because of its low water table. Leaders also found out that the landfill had been chosen because two additional waste projects were also planned for the same area. A 1983 report by the General Accounting Office revealed that three out of four of the off-site commercial hazardous waste landfills in eight southern states happened to be located in predominantly black communities, even though blacks made up only 20 percent of the region's population (U.S. General Accounting Office, 1983).

The incident created an entirely new area of legal practice called environmental justice, described previously in Chapter 1. It has also become known as a watershed event in contemporary environmental activism and the birthplace of the environmental justice movement.

Because the protests had taken place during an election year, those same participants became political activists, and voter registration in Warren County increased by 30 percent; most of the new voters were nonwhite (McGurty, 2000). This large increase in voter registration, especially among black residents, had a profound effect on county politics. In November 1982, races for the majority of county officials, including the county sheriff, were won by black candidates.

Sources and Further Reading

Bullard, Robert D., ed. *Confronting Environmental Racism: Voices from the Grassroots.* Boston: South End Press, 1993.

Bullard, Robert D., and Glenn S. Johnson. "Environmental Justice: Grassroots Activism and Its Impacts on Public Policy Decision Making." *Journal of Social Issues* 56, no. 3 (Fall 2000): 555–578.

Gottlieb, Robert. *Forcing the Spring: The Transformation of the American Environmental Movement.* Washington, DC: Island Press, 1993.

McGurty, Eileen M. "Warren County, NC and the Emergence of the Environmental Justice Movement." *Society and Natural Resources* 13, no. 4 (June 2000): 373–388.

United Church of Christ, Commission on Racial Justice. *Toxic Waste and Race in the United States.* New York: United Church of Christ, 1987.

U.S. General Accounting Office. *Siting of Hazardous Waste Landfills and Their Correlation with Racial and Economic Status of Surrounding Communities.* Washington, DC: GAO, 1983.

Land of the Fee?

Environmental activists in several states across the United States have banded together in a somewhat unusual coalition that includes hikers, off-highway vehicle users, photographers, boaters, tourists, skiers, horseback riders, and rock climbers. The issue? Whether access to public lands ought to be free, or whether entrance should be limited by a fee.

In 1996, Congress authorized the Recreational Fee Demonstration Project, which allows the U.S. Forest Service, the National Park Service, the Bureau of Land Management, and the U.S. Fish and Wildlife Service to implement user fees at locations across the country. The original intent of the program was to charge entrance fees that would be reinvested for maintenance and improvement of recreation facilities and services. In the wake of large-scale budget cuts during the 1990s, coupled with increased recreational use, federal agencies had found they could not maintain trails, provide rest room facility and parking lot maintenance, or staff sufficient levels of law enforcement. The fees, which were levied at 376 locations in 2000, range from $2 to $5 per day, per vehicle, to $5 per person, with options for purchasing monthly or yearly passes. Some areas that previously had no fees for services or activities began charging for the first time.

In congressional testimony in 1999, federal officials said that the fee demonstration was quite successful and that public acceptance was increasing. Although there was initially some opposition, surveys, public comment cards, and newspaper coverage showed that people grew more accustomed to paying fees over time. They were also more likely to support the fee program if they saw obvious, quick improvements at the site and if the fees were easy to pay. By law, the majority of the recreation fees are retained by the agency that collects them—an incentive, some believe, for excessive enforcement.

Despite agency claims, there has been a rising revolt over what some call the "pay to play" program. An estimated 200 grassroots groups (*Arizona No Fee News*, 2001, 4) across the country have organized where the fee demonstration project has been implemented, with the Western Slope No Fee Coalition one of the strongest protest groups. In July 2001, the organizers protested at a new fee area in the San Juan Mountains of Colorado. They were met with two roadblocks and armed forest rangers who gave the protesters the option to leave. Fifty members of the No Fee Coalition proceeded into the fee area and patiently waited in line to receive a $25 federal citation. Other protesters who refused to buy passes faced the possibility of six months in jail and a $5,000 fine. Even though there is little question that the fee program is entirely legal, several activists have chosen to go to court to protest the fees, with one case reaching the U.S. District Court in

Oregon. Even local officials have protested against the fees, which they believe are taking away local control.

Critics of the fee demonstration program raise several objections that are similar to those of members of the wise use movement who argue that the public lands should be open to all forms of public use. Some protesters argue that too much of the collected money goes to administrative overhead rather than directly to area improvements. One group, Wild Wilderness, believes that the fees collected in the national forests are a mask for substantially lowering appropriations in the wilderness budget. They seek a clear added value to the public when fees are charged.

Another concern relates to equity and whether or not the fees discriminate against working-class members of society. They note that wealthy visitors can easily afford the fees and that the poor are not likely to come anyway. When Glacier National Park first started collecting weekend fees in the winter of 1997–1998, the National Park Service announced that it had collected over $3,500. But it did not report the number of people who turned away at the park entrance because they did not want to pay the fee. Local newspapers claimed that about 100 vehicles, or possibly 250 individuals, had ended their visit because of the fee (Matthews, 2000, 7).

The U.S. Forest Service contends that because the fee program is still in the demonstration phase, part of the project is to learn how to minimize the impact on people with low or fixed incomes. There are different pricing options, from a daily pass to an annual pass that provides substantial savings over the course of a year. In most areas where the fee program is in effect, visitors can exchange volunteer work for a free volunteer pass.

Several Forest Service fee demonstration programs have been evaluated and altered as a result of the opposition of protest groups. In the White Mountains of New Hampshire, and in Idaho's Sawtooth Range, fees have gone from being required anywhere a vehicle is parked to being required in specified locations that are heavily used. Fee structures in Oregon and Washington were changed when the public was faced with paying multiple fees required by federal and state agencies, often paying two fees for a single recreational outing.

In northern Arizona, opponents of the Red Rock Country pass have engaged in a unique form of civil disobedience. They created their own rearview mirror tags that claim exemption to the fees, noting "The occupants of this vehicle are not engaged in a recreational activity." The Arizona No Fee Coalition claims that the fees are double taxation and economic discrimination and that most of the money is being spent on enforcing the pass regulations. During its first year of operation, the Red Rock Country project raised nearly $600,000, and officials admit that the fee program was only able to break even that year. Critics argue that for the fiscal year 2000, only 4 percent of the $25.4 million collected nationwide was spent on habitat enhancement and resource preservation, whereas 52 percent went to collection, operations, and law enforcement (Clifford, 2001). One member of Congress whose constituents opposed fees in her district sponsored the Forest Tax Relief Act of 1999, which would have removed the Forest Service from the fee demonstration program. State legislatures in Oregon, California, and New Hampshire have passed resolutions that oppose the program.

Agency officials respond that recreational use of wild areas is rapidly increasing and that the government is required to meet the needs of a growing number of senior citizens and persons with disabilities. An emphasis on environmental education calls for more interpretative programs and signage, and health and safety standards must be met if areas are not to be closed down. Administrative costs are being reduced as the program gains more experience and becomes more efficient at fee collection. Accommodations have been made for free use by school groups or for American Indian tribes who use the forest for cultural or religious purposes.

Those who oppose the fee demonstration program fear that it will go into effect nationwide at some point, even in remote areas with few visitors. Other support a lobbying effort to make sure that Congress provides resource agencies with sufficient funding to make sure recreational use is made as important as resource extraction.

Sources and Further Reading

Arizona No Fee News. Flagstaff, AZ: Arizona No Fee Coalition, Summer 2001.

Bates, Nicole A. "Paying to Play: The Future of Recreation Fees." *Parks and Recreation* 34, no. 7 (July 1999): 46–52.

Clifford, Hal. "In the Great Outdoors, Resistance to Rising Fees." *Christian Science Monitor,* 21 August 2001.

Free Our Forests. http://www.freeourforests.org.

Wild Wilderness. http://www.wildwilderness.org.

Matthews, Mark. "Working Class Can't Foot the Bill." *High Country News,* 14 February 2000.

McManus, Reed. "Land of the Fee." *Sierra* 84, no. 6 (November–December 1999): 22–23.

Oriol, Bill. "Tagging a Protest." *High Country News,* 29 January 2001.

Monumental Decisions

On 8 June 1906, President Theodore Roosevelt signed the Antiquities Act, the purpose of which was to protect archaeological sites on public and Indian lands owned by the federal government. The statute gave the president the power to set aside historic sites as well as natural and scientific areas that are considered to be especially significant. It took environmental activists twenty-five years to gain sufficient support for the measure, in part owing to concerns about the plundering and destruction of archaeological sites and places sacred to Native Americans.

Efforts to preserve wild and historic areas began in 1882 with Massachusetts senator George F. Hoar and the continuing struggle between conservationists and preservationists. The conservationists' concern was whether or not the federal government had the resources to protect areas, not whether they ought to be protected in the first place. Preservationists pointed to the vandalism and looting that were already taking place at such sites as Chaco Canyon and Mesa Verde as a rationale for specific protective legislation. In 1891, President Benjamin Harrison signed the Forest Reserve Act giving the president the right to set aside land for government management, and two years later, he designated 13 million acres as protected lands ("Cultural Landmarks Revisited, 2001"). By the turn of the century, Progressives had made some progress in convincing Congress that federal intervention was necessary and that there would be a substantial public benefit that went beyond the commercial acquisition of resources at any given site. In 1901, government possession of public lands increased to 46 million acres under Presidents Grover Cleveland and William McKinley ("Cultural Landmarks Revisited, 2001"). John Muir visited the Petrified Forest in 1905 and was angered by railroad personnel who loaded petrified logs onto trains, later to be made into tourist trinkets. It was largely at Muir's suggestion that Roosevelt created the Petrified Forest National Monument.

There was a clear distinction at that point between the national monuments, which were assumed to have scenic, rather than commercial, value, and the national parks. Monuments included areas of scientific curiosity and those that were made by humans, combined with the requirement that they be confined to the smallest area possible. One of the first national monuments was designated in what was then the New Mexico Territory. Roosevelt proclaimed the 160 acres of El Morro and Inscription Rock as of the greatest historical value. In 1908, he proclaimed Muir Woods National Monument in California, setting a precedent by protecting land that had been privately owned and a gift from a California couple. In Wyoming, Devil's Tower, a spiral basalt formation, was included, as was Arizona's Montezuma Castle, a cliff dwelling protected because it had been built by humans. Roosevelt stretched the definition by initially declaring the Grand Canyon as a monument, despite its size and the fact that it was still part of the Arizona Territory.

Between 1906, when the Antiquities Act was signed, and 1978, the statute was invoked by a dozen presidents to declare ninety-nine national monuments; of these, thirty-eight were considered historic or prehistoric and sixty-one areas were primarily scenic and natural ("Cultural Landmarks Revisited, 2001"). In south-central Ohio, the prehistoric Hopewell Mounds near the Camp Sherman military reservation gained the protection of President Warren G. Harding. The fifty-seven-acre parcel along the Scioto River had been singled out for protection by the Chillicothe, Ohio, Rotary Club, which worked with Ohio's political officials to secure support. The Ohio State Archaeological and Historical Society was later granted a revocable license to preserve and protect the monument—an important step because most sites were under the care of volunteers.

By 1990, there were fifty-two monuments, as several of the areas became part of a national park; others were made into national historic parks, a historic battlefield, and a national historic site; two became a national parkway; and eleven were abolished. The largest withdrawal (lands withdrawn from the public domain and set aside for environmental protection by the government) came during the administration of President Jimmy Carter, who created 56 million acres of national monuments in Alaska in 1978, even though the land was not in any danger of being damaged.

President Clinton gained the support of environmental activists, and the ire of numerous other interests, when he used the Antiquities Act to designate 5.7 million acres as national monuments, starting in January 2000 and continuing throughout his last weeks in office. The new sites that gained federal protection ranged from the Canyon of the Ancients, west of Colorado's existing Mesa Verde National Park, to the U.S. Virgin Island Coral Reef National Monument, an area said to contain all the elements of a Caribbean tropical marine ecosystem. Over 1 million acres were designated during his last week in office, with critics arguing that he did so without consulting elected officials and in defiance of the purpose of the Antiquities Act.

This is where the crux of the problem lies. How can areas of historic, scenic, and natural beauty be protected without overstepping presidential discretion under the Antiquities Act or without congressional consultation? Clinton interpreted his authority very broadly, acting often at the behest of specific environmental organizations who pled their cause before Interior Secretary Bruce Babbitt. In doing so, he was accused by one member of Congress as "acting more like an emperor than President of the United States" (Nickles, 2001). Although Theodore Roosevelt designated 1.5 million acres under the Antiquities Act and Franklin Roosevelt nearly twice that amount, Dwight D. Eisenhower removed 22,500 acres from protection. Richard Nixon, Ronald Reagan, and George H. W. Bush made no designations during their administrations.

After Clinton left office, resource extraction advocates, recreationists, and industries attempted to have the Antiquities Act designations overturned. In California, Sierra Forest Products, the Sierra Nevada Multiple-Use and Stewardship Coalition, Tulare County, and other groups challenged the legality of the designation of the Giant Sequoia National Monument, created 15 April 2000. The monument designation would preserve thirty-four of the last seventy groves of giant sequoia, the largest trees on earth. The 327,769-acre area identified by the Clinton administration was designed to protect not only the trees themselves but also the grove's watershed and ecosystem. Supporters, including Earthjustice, the Natural Resources Defense Council, the Sierra Club, the Wilderness Society, and the Tule River Conservancy, argued that the monument designation was the first management program that recognized the relationship between the physical environment and living organisms. The plaintiffs also challenged the designation on the basis of provisions in the National Forest Management Act, the National Environmental Policy Act, and the Administrative Procedure Act. A federal judge in Washington, D.C., threw out the lawsuit in October 2001, reaffirming the president's power. The ruling was the seventh in a line of other legal challenges, and each time, the courts upheld the president's use of the Antiquities Act.

Sources and Further Reading

"Court Resoundingly Rejects Challenge to Giant Sequoia National Monument" (2 October 2001 press release). http://www.earthjustice. org/news. Accessed 5 October 2001.

"Cultural Landmarks Revisited: The 1890 Census and the Roosevelt Presidency." http://www.xroads.virginia.edu/~CAP/NPS. Accessed 24 April 2001.

National Park Service. "The Proclamation of National Monuments under the Antiquities Act, 1906–1970." http://www.cr.nps.gov/aad/pubs. Accessed 29 April 2001.

Nickles, Senator Don. "President Clinton's Use of 1906 Antiquities Act." (22 January 2001 Remarks at U.S. Senate). http://www.reagan.com/ HotTopics. Accessed 25 April 2001.

The Reagan Information Interchange. "President Clinton's Use of 1906 Antiquities Act" (22 January 2001). http://www.reagan.com/HotTopics. main. Accessed 25 April 2001.

U.S. General Accounting Office. *Federal Land Management: Information on Usage of the Antiquities Act.* Washington, DC: U.S. General Accounting Office, 1999.

Wilderness Society. "America's National Monuments." http://www. wilderness.org/standbyland/monument. Accessed 24 April 2001.

3

Global Environmental Activism

lthough the focus of the book thus far has been on environmental activism in the United States, it is also important to analyze how that movement has influenced people and policy-making in other countries. Problems such as air quality and water pollution, global warming, nuclear waste, and energy resources are experienced globally, and as a result, individuals and organizations have coalesced around environmental issues worldwide. The term often used to describe the elements of these movements is *non-governmental organizations* (NGOs). This differentiates the activism of grassroots organizations from the official policymaking conducted by governments and their leaders.

Global environmental activism is frequently linked to political and economic events, especially in newly developing areas. One study has found that as education and per capita income increase, environmental quality initially worsens; but at some point, this worsening peaks and then declines (Desai, 1998). Similarly, a growth in pollution-intensive industries and a lack of environmental regulation eventually evolve into service- and information-intensive activities, reducing the burden on the environment. The result is a gradual change in public opinion that demonstrates that there is a serious concern about environmental problems, although primarily those at the local level.

NGOs associated with environmental issues play a significant role in the global debate. They tend to serve as outspoken critics of the government and transnational corporations, highlighting activities that degrade the environment. They often serve as the independent monitors of international agreements and as linkages to international governmental organizations such as the

United Nations. They may provide a forum that builds public awareness about environmental problems and, in some cases, become parties in legal proceedings against nations that fail to comply with international treaties and regimes. They have been less successful in affecting the actions of the World Bank and World Trade Organization, both of which finance large projects in developing countries.

This chapter reviews environmental activism by looking at each of the earth's regions and subcontinents as a way of understanding the similarities and differences among organizational types, issues, and strategies. It is important to note that on a global level, NGOs are changing rapidly in response to fluctuations in economies, political leadership, and sociocultural events. Unlike the situation in the United States, where environmental activism appears to have settled into a somewhat predictable pattern of tactics and goals, other nations are experiencing new challenges and opportunities that sometimes follow the pattern of the United States but also mirror their own priorities and culture. The chapter concludes with a look at international environmental organizations that have shaped policies through their activism over the past two decades.

Africa

Dr. Wangari Maathai stands as the model for African environmental activism—the first woman to earn a doctoral degree in eastern and central Africa and the founder of Kenya's Green Belt Movement in 1977. A lecturer in anatomy at Nairobi University, Maathai was originally involved in a fight against poverty led by the National Council of Women of Kenya. She started a grassroots movement of women by creating a small tree nursery in her backyard, aiming to curb soil erosion and also to demonstrate the need for sustainable fuel use among her people. Today, the Green Belt Movement includes over 5,000 nurseries throughout Kenya, providing an income for 80,000 people and resulting in the planting of over 20 million trees ("Green Belt Movement," 2002). The project has been expanded to thirty countries in Africa, but grassroots activism has been slow or nonexistent.

The pace of activism varies considerably on the African continent, for several reasons. Many countries are consumed by tribal and ethnic unrest that forces any other policy issues out of the way. Only in those areas where there has been a significant amount of urban growth are issues such as pollution and energy being widely addressed. Activism is of a much more recent nature, with most of the NGOs having been formed in the 1990s. Other differences occur in countries that were the subject of imposed colonialism, primarily by the British, French, and Italians, or through Cold War imperialism. When areas were partitioned, there was often a concurrent military buildup and the development of corrupt political elites whose policy agenda did not include the environment.

In Somalia, for instance, an estimated 400,000 people have been killed as a result of recent clan-oriented civil wars and war-induced famine and disease. Almost half of the country's population has been displaced—1 million are believed to have fled to neighboring countries and the Middle East (Adam, 2001). Despite the fact that the country is situated on the heaviest oil tanker route in the world, the military government has had virtually no oil pollution–fighting equipment or staff. Somalia has lacked such basic state institutions as a legal system, telephone and postal networks, education and health systems, and electricity and water providers. The result is a region beset by the exploitation of marine resources, a dumping ground for hazardous wastes, and the challenge of restoring some form of political structure.

The overwhelming barrier, however, is poverty. The World Bank has estimated that the fifty nations of sub-Saharan Africa (which excludes South Africa), with a combined population more than twice that of the United States, have a combined gross domestic product of less than $200 billion, roughly that of Virginia (Hayes, 2001). There simply is insufficient internal capital to finance or prioritize the protection of natural resources and the continent's air and water.

The exception is South Africa, where the constitution specifically guarantees the rights of citizens to live in a clean and healthy environment. Publicly, at least, the country's environmental minister has sought to place a greater emphasis on the problems of pollution and waste. Following the first multiracial elections in 1994, grassroots groups called upon the new government to enact stricter environmental laws, along with a framework for dealing with air, water, and marine pollution; deforestation; diversity; and energy efficiency. Despite their efforts, the country does not have any legally binding air pollution control regulations, and the government lacks enforcement capabilities. Despite a June 2000 oil spill from a tanker rounding the South African coast that damaged tourist areas and the local penguin population, activists have charged that the government still lacks a comprehensive plan to deal with marine pollution.

But energy has been the dominant environmental issue, since deforestation and rapid population growth and urbanization have accelerated since the mid-1990s. Since 1970, South Africa has consumed the most energy among the major countries of Africa, even though about one-fifth of the rural population does not have access to grid electricity. In rural areas, an estimated 3 million people burn wood to heat their homes and for cooking. This leads not only to air pollution but also to respiratory problems, since the wood is often burned in small enclosed spaces without ventilation. The appeal and ready availability of low-cost coal, a lack of public awareness about energy costs, and the absence of regulations are cited as the reasons why energy efficiency measures have not been adopted.

Some NGOs have been successful in bringing change to the country. In 1999, the Save the Vaal Environment (SAVE) group successfully petitioned the government to stop the development of a strip mine on the banks of the Vaal River. The parties were brought together in a forum to discuss their differing perspectives and reached some agreement on the utilization of the region's natural resources. But the majority of the advances in South Africa's environmental policies have come because of the influence and support of international groups and the United Nations, not from within its own people.

In central Africa, especially in the postindependence countries such as Cameroon, Gabon, the Congo, and the Central African Republic, a gap between the rulers and the ruled has made environmental activism almost nonexistent. So-called prestige projects (mining, highway construction, railways, oil, and forestry) are controlled by elites with close ties to the government. Despite the differences in their colonial histories, these nations have in common strong forces that encourage graft and corruption, frustrating the efforts of the few groups that attempt to oppose them. Some indigenous groups have attempted to monitor international agreements such as the International Tropical Timber Agreement (ITTA) and the Convention of International Trade in Endangered Species of Wild Flora and

Fauna (CITES), but their resources are very limited. They often rely upon university researchers and global organizations for scientific expertise.

Nigeria is typical of much of Africa's struggle to build environmental coalitions with any real influence. The problems in the country are not unique: drought, soil erosion, desertification, deforestation and degradation of vegetation, pollution, and waste dumping. But the country also has an early history of activism, beginning with the founding of the Nigerian Field Society in 1930. Today, however, the power and influence of the Nigerian government pervade all attempts to bring the environment to the political agenda. NGOs are simply not in a position to accomplish much without governmental support, with some activists believing the relationship between groups and government officials is much too close. For example, in 1991, the government unilaterally selected one of the NGOs to be the Nigerian Environmental Society (NES) as the guardian of the environment. The coziness between the NES and officials has led to the creation of a new acronym: GONGO, or government-owned nongovernmental organization. Charitable groups such as the Nigerian Conservation Foundation work more on environmental education than on policy development and education. Still, as in many African countries, environmental activism is focused on the national level, rather than through grassroots groups.

Polar Regions

Protection of the natural resources in Antarctica and the Arctic regions has been accomplished through a variety of strategies that are somewhat different than those found in other regions. Antarctica, for example, is unique because several nations claim sovereignty over the continent, and there are no indigenous people to fight for its protection. As a result, attempts to preserve its pristine environment and huge store of natural resources and wildlife have been coordinated by global nongovernmental organizations. Perhaps the largest is the Antarctic and Southern Oceans Coalition (ASOC), founded in 1978 by the Friends of the Earth International, the World Wildlife Fund, and several other groups dedicated to the inclusion of the public in decision making in meetings of the Antarctic Treaty System. The ASOC now includes about 240 organizations in fifty countries ("Antarctica,"

2002). The groups have engaged in lobbying activities to persuade participants to fully comply with the Environmental Protocol to the Antarctic Treaty and to stop illegal fishing in the Southern Ocean.

Throughout the 1970s and 1980s, global groups initiated various plans to create an Antarctica World Park. One proposal would have erected an invisible, but legal "fence" around the entire continent as a way of eliminating all human activity. Other organizations sought to dissolve all territorial claims, whereas still others supported the idea that all nations would have a share in the continent's resources and benefits. By the 1990s, the emphasis had slightly shifted to the issue of scientific exploration and tourism. Although most researchers agreed that cooperative study was a positive step, environmental groups argued that private tour operators not associated with any national Antarctic program were incompatible with the idea of preservation. Greenpeace, one of the leading activist NGOs in Antarctic matters, has proposed that the continent be maintained as a peaceful, demilitarized, nuclear-free region, with its wilderness qualities preserved. All activities would be judged by their potential for conservation, with science the primary reason for any human activities. Last, Greenpeace proposes a ban on the storage, discharge, or incineration of toxic or radioactive waste or the use of nuclear reactors for any purpose.

Antarctica's fate lies in the hands of seven governments who have asserted claims to the continent since 1907. Since those claims are not recognized by the international community, environmental activists must rely upon a number of regimes, beginning with the 1959 Antarctic Treaty, and the twenty-six governments that now participate as consultative parties, for the continent's protection.

Environmental activism in the Arctic regions of the world covers a broad swath of geography and issues and thus involves organizations and indigenous peoples in the eight Arctic rim nations. Activism and political participation in these regions is much more recent, based in part on the newly organized groups that have formed around the interest of native peoples.

There are some organizations that seek to coordinate activism in arctic areas, such as the Washington, D.C.–based Circumpolar Conservation Union. Its environmental research and coalition-building activities represent the combined efforts of groups concerned about both pollution and sustainability as well as human rights. Similarly, global NGOs such as Greenpeace and the Natural Resources Defense Council have become stakeholders in protecting valuable natural resources. Although the Arctic regions were at one time pristine, they are now subject to acidifying pollutants from Russia and Eastern Europe, persistent organic pollutants (POPs) such as pesticides, oil pollution from shipping, and radioactive dumping.

Many groups focus on providing assistance and resources to native people who have historically been less organized in confronting the increasing encroachment of resource extraction. Exemplary of these efforts is the fight by activists to preserve the Arctic National Wildlife Refuge from oil drilling or other development. In the late 1980s, NGOs began to work with indigenous groups such as the Gwich'in Steering Committee to enlist them in the protection of the area's water, air, and, especially, the caribou population. The Gwich'in Nation has subsequently received support from Indigenous Survival International, the National Congress of American Indians, and the International Indian Treaty Council.

Not all of the indigenous groups are in agreement as to what kind of environmental protection is needed, especially in the case of marine mammals. The High North Alliance, based in Norway, has supported the right of the people living in coastal communities in Arctic areas to use marine mammals sustainably. The group also monitors whaling by such countries as Iceland, Japan, and Norway that have frequently been the targets of environmental protests by Greenpeace and other organizations. Since the International Whaling Commission has no enforcement powers over nations that violate international agreements on commercial whaling, NGOs become the primary monitors over the whaling industry.

The boreal forests of Siberia and North America provide another battleground for activists seeking to protect old-growth forests and the people who live there. The Taiga Rescue Network (TRN), for instance, has documented the cultural losses of the Evenks—an indigenous people of Siberia and the Russian Far East. Many of the clans lost their national identity and language as development descended upon their villages in the early twentieth century. Similar concerns have been raised about the forests of Norway, Sweden, and Finland, whose old-growth trees became prime candidates for logging after environmental groups in the United States were successful in their efforts to change timber companies' policies.

Asia

Reliance upon international groups for support is not uncommon in Asia, where environmental awareness is still in the developmental stage. The World Wildlife Fund (WWF), for example, has prioritized the protection of coral reefs, beaches, and marine life in the Sulu Sea, which surrounds the Philippines, Indonesia, and Malaysia. Overfishing and the use of cyanide and dynamite for capturing rare tropical fish for collectors have harmed the marine ecosystem and depleted fish populations. The WWF used its extensive resources to set up community enforcement efforts, funding citizen patrols to apprehend poachers in national parks and reserves.

Asian environmental activism differs considerably from one nation to another, especially between industrialized and newly developing economies that are undergoing a transition to democracy. In South Korea and Taiwan, for example, a shift from authoritarian rule has been accompanied by a change in the role of the media, which are now able to report on industrial pollution, often serving as advocates for government cleanup projects. There has been a parallel shift in emphasis from small groups focusing on local environmental problems to national NGOs that engage in traditional political strategies such as lobbying public officials, supporting political parties, and creating a substantial membership base. In some countries, it is now routine for polluters to compensate citizens for damages the polluters have caused to crops or health.

One exception to the pattern is Thailand, where the shift to a more democratic system has been reflected in the creation of elites who are more likely to reflect the concerns of business interests and where vote buying is common. There, citizen groups have but a small voice in efforts to improve the environment, and the government has little support for changes that would regulate industrial and economic growth.

Democracy itself does not ensure that government will prioritize the environment; even in democratic governments such as that in the Philippines, the vocal environmental protest groups have had little impact. Instead, less contentious groups such as the Haribon Foundation have focused on public education and awareness rather than direct action. The foundation, created in 1972, began as a bird-watching group and in 1983 became a conservation foundation conducting research on biodiversity.

In semidemocratic Malaysia, however, public pressure to control pollution resulted in the creation of an effective and powerful national environmental protection agency. Singapore, in contrast, changed its policies because of leadership from the upper tiers of government, not because of grassroots activism.

Groups in other parts of Asia have been more successful (and have a longer history) in establishing and gaining public support. In 1982, a group of citizens banded together in India to try to protect one of the nation's holiest sites—the River Ganga (Ganges). Meeting at a temple site in Varanasi (Banaras), the most holy of Indian cities, religious leaders, engineers, and others concerned about the heavily polluted river formed the Sankat Mochan Foundation (SMF).

Women have always been a key element of Indian activism, starting with the *chipko* movement against logging that was led by women. One current environmental leader, Vandada Siva, is a physicist who has supported small farmers against corporations that are attempting to control the production of native plant species such as rice. She heads the Research Foundation for Science, Technology, and Natural Resource Policy in Delhi, and her books have influenced international policy related to dams and agriculture.

Japanese environmental activism is a recent phenomenon, stalled by the push for economic growth and industrialization that has taken place in the country. As early as the 1930s and 1940s, the militarized economy led to massive deforestation and pollution, with little attention to environmental problems. Just after World War II, scientists found that rice had been tainted by cadmium, leading to major outbreaks of industrial poisoning. The political system's emphasis on rapid growth, large-scale land development, and consumption far outweighed environmental interests through the next three decades, even when the country faced international criticism for its policies on whaling and timber. The issue of the environment did not reappear until four very serious pollution incidents were exposed and lawsuits were filed by victims of water and air pollution between 1967 and 1969. By 1973, the courts had decided for the plaintiffs in all of the Big Four cases, and the Japanese media triggered public and political support for environmental protection. At that point, community-based activism slowly began to emerge.

But the momentum that the Big Four cases started appears to have waned over the past three decades. Today, Japanese environmental activism often comes from young leaders who have been trained in the United States. University students frequently model their activities on U.S. organizations, and one Japanese group, Global Eyes, founded in 1997, is attempting to develop environmental education programs, stimulate recycling efforts, and conduct environmental research. Even though "educational" activities are considered appropriate, litigation and protest are somewhat of an anathema to Japanese culture, which seeks to minimize conflict in place of mutual agreement and understanding. But some researchers argue that this may also be a result of the fact that the Japanese government has been successful in eliminating or at least reducing the pollution that originally led to the 1960s litigation and that litigation is time consuming and costly. There is a scarcity of judges in the country to resolve disputes, along with a lack of trained attorneys.

The powerful military presence and past policies of the cultural revolution negated almost all attention to the environment in China. Natural sites were under local or military control, with devastation the result. Political and economic reform in China in 1978 forced new leaders to deal with environmental issues after years of neglect. The government's increasing involvement, especially in the past two decades, is likely due to the desire to expand international ties and generate hard currency rather than to any growing environmental awareness on the part of citizen groups, however. The governmental regime is still largely authoritarian, preventing the development of strong NGOs or debate among conservationists or within the media. One study found that most Chinese citizens are unaware of the dangers of air or water pollution and have even less awareness about global issues such as biodiversity or climate change. The few linkages that have been created with the West have come largely through visits by foreign scholars and cultural exchanges, mostly in the research community in Beijing.

Although it is somewhat of a generalization, the sheer geographic size of the country and its population mean that the farther one gets from the capital, organized activism is almost nonexistent. The Chinese government's myriad bureaucracies control virtually every aspect of the environment, from the export of pandas to ocean dumping. International groups such as the World Wildlife Fund have been hesitant to commit more dollars for species protection, for example, after finding that funds destined for panda protection at the Wolong reserve actually were spent to build a hotel and school for the children of resident Chinese scientists. The NGOs' reluctance is built upon similar instances that indicate a lack of public support for stopping illegal trade in endangered species or overfishing.

Europe

Although considered a single continent, there is still a wide gap between the eastern and western regions of Europe. In the east, democracy is young, and environmental protection has traditionally been low on the policy agenda. Massive amounts of military spending during the Cold War and into the 1980s have left a wide swath of degradation. In the west, nations are more industrialized but also subject to transboundary environmental problems that demand cooperation. As a result, activism varies tremendously between the east and west, as the summaries below indicate.

Eastern Europe

Activists in the republics of the former Soviet Union are considerably different from those in other nations in their focus and power. For decades, there was little environmental awareness among Soviet citizens because of the secrecy surrounding the government's actions and projects. Only in the past ten years has there been much acknowledgment of the magnitude of the problems of pollution, much of it military waste that was dumped or left untreated.

The Russian Federation has increasingly found itself involved in environmental issues whether it wants to be or not. At one time, the former Soviet Union manufactured about one-eighth of the world's total of ozone-depleting substances. During the regime of Mikhail Gorbachev, who became general secretary of the Soviet Communist Party in 1985, the Soviets realized the need to play a role in international agreements that directly affected their economy. They became one of the first governments to take steps to protect the ozone layer and to prevent transboundary air pollution. The policy of glasnost—political openness—unveiled an environmental history that began to show the enormity of Eastern Europe's pollution problems, but it also opened up some government data that had previously been kept secret from the Russian public. Although pre-Gorbachev reporting and record keeping was problematic, the 1991 collapse of the Soviet Union at least gave researchers some idea of the problems the region faces.

Even though Russian scientists and researchers have attempted to build activist coalitions, they rarely have the necessary resources to make a real impact. Instead, business groups such as Ecological International, headquartered in Moscow, have been established to create an international fund to protect the ozone layer. Although the group's title might appear to be pro-environmental, it is noteworthy that the support for the fund came from an industrial association and enterprises of the Ministry of Defense.

The protection of wildlife in Eastern Europe has been coordinated by U.S. or global organizations rather than indigenous groups. The National Geographic Society and the National Fish and Wildlife Foundation, for instance, provided grants to sustain the Siberian tiger, as did the Exxon Corporation, whose mascot is a tiger. Greenpeace International has actively pressured the government to halt ocean dumping, and in the late 1980s, the Natural Resources Defense Council signed an accord with the Soviet Academy of Sciences to conduct research on energy efficiency in buildings and appliances to reduce ozone-depleting gases.

Realizing their own resource limitations, several Russian conservation groups made an appeal to President Bill Clinton for support for the imposition of economic sanctions on China and Taiwan in order to stop the poaching of endangered wildlife. The Far Eastern Leopard Club, the International Clearinghouse of the Russian Far East, the Khabarovsk Wild Life Foundation, and other groups realized that their own government was not sufficiently motivated to ensure compliance with international treaties and protocols.

Most NGOs are small and local, and even though Russian environmentalists are more organized than ever before, many of the smaller, single-issue groups disband after a successful or unsuccessful confrontation with the Russian government. Environmental concerns are now low on the public's agenda, pushed aside by such issues as public health and the economy. The Forest Club, an informal working group composed of representatives from the largest NGOs, has attempted to improve Russian forest legislation and regulations, conduct an inventory of natural old-growth forests, and monitor protected territories for illegal poaching and logging. Members include the Socio-Ecological Union's Forest campaign, the Biodiversity Conservation Center, and the Student Corps (Druzhinas) for Nature Conservation.

The Russian government itself has sent mixed messages to the rest of the world regarding its interest in environmental protection. In 2002, the Russians agreed to import 20,000 tons of high-level radioactive waste from other countries. Fearing that the country was about to become an international nuclear waste dump, groups organized a series of protest actions, including an antinuclear walk and an action camp, in June and July 2002. The protest was organized by three groups—Ecodefense, the Socio-Ecological Union, and Greenpeace-Russia—near the Siberian city of Krasnoyarsk. Despite a petition from 100,000 Krasnoyarsk residents seeking a vote on a regional ban on nuclear waste imports, local government officials refused to recognize the signatures and the referendum process, which is part of the Russian constitution. The environmental organizations conducted training on nonviolent protests for people who have only recently been allowed to engage in such strategies.

Despite the Hungarian government's participation in several international environmental accords, Hungary's citizens have developed little awareness of environmental problems that plague their country, including the lack of preservation plans for natural areas and severe water and air pollution. Instead, the government has appeared to treat support for environmental protection as a means of gaining favor with western countries. What little concern for the environment there is comes from mid-level bureaucrats within the Hungarian government, rather than from political leaders or citizen groups. Only one major issue appears to have mobilized activism—opposition to the Gabcikovo-Nagymaros Dam—which is believed to have been one of the factors leading to the demise of the socialist regime in the late 1980s.

Western Europe

Western European environmental movements have often paralleled activities in the United States, although there are some differences in the targets of their activism. Historically, Europeans belong to some of the strongest antinuclear groups in the world, and although their focus may have shifted slightly in the past twenty years, their opposition has not weakened. As early as 1980, activists throughout the continent began to monitor and oppose the transport of spent nuclear fuel, much of which is shipped from Japan and from other European countries to Britain. Organizations such as Cumbrians Opposed to a Radioactive

Environment (CORE) have conducted protests against the rail transport of spent fuel from power stations and submarine reactors to the Sellafield reprocessing plants in the United Kingdom.

Great Britain is a remarkable case study in environmental activism for several reasons that distinguish the country from others. First, it has a lengthy history of grassroots activity that has often been supported by the royal family. Second, the nation's activists have been involved in a broad spectrum of issues not unlike those of American groups. Third, activists are working hard to eliminate the nation's singular label as the Dirty Man of Europe. Fourth, British activists are known for direct action, especially protests and ecoterrorism, but they have also recently turned to Internet activism.

The most recent examples of direct action, which started around 1997, are related to the controversy over genetic engineering of plants and animals. Companies such as Monsanto and Novartis, along with academic institutions involved in plant research (such as Leeds University), have been "visited" by crop saboteurs with such names as Captain Chromosome, the Lincolnshire Loppers, the Gaelic Earth Liberation Front, and GenetiX Snowball. Their tactics include uprooting fields where genetically modified crops have been planted and ransacking seed storehouses, and many use organizing strategies closely allied with the antinuclear and antiroads movements.

There has also been a proliferation of groups operating at the local level whose concerns are specific to their community. Typical are organizations such as Defend the Dunes, established when a proposal was made to construct twenty-six homes on a greenbelt that is part of a Site of Special Scientific Interest (SSSI). The group is working to stop all development on sites in the area and to expand the greenbelt.

Some groups are similar to, but not necessarily modeled after, an American group. In the United States, for example, the Surfrider Foundation was initially established by surfing enthusiasts who sought to protect the California coastline from increasing pollution. Surfers Against Sewage, which was founded in the United Kingdom in 1990, has a similar agenda, campaigning to clean up beaches and rivers to provide recreational users with a safe water environment.

British promotion of direct democracy is somewhat contradicted by environmental groups' use of the Internet, since the production and usage of computers have extensive environmental consequences. One study found that activists' use of electronic technology tended to decentralize organizations and prevented the development of hierarchical organizational structures. The Internet also serves as a useful tool for mobilizing both group members and the public when there is an important issue at stake. Networking among groups, who are often fragmented and uncommunicative, is also enhanced and creates a sense of solidarity among activists. In addition to serving as a channel for information, some activist groups in Britain use the Internet for what is called "hacktivism," gaining access to adversaries' computer systems as a form of protest. Overall, Internet use appears to be a strategy that is one more tool in activists' repertoire rather than representing a shift in types of protest methods.

As an advanced, industrialized country, Germany has long been perceived as Europe's Green Knight because of the attention paid to environmental problems. In part, this perception is due to the degradation and pollution that occurred over the prior fifty years. The country's rivers experienced massive pollution, air quality suffered from industrialization, and a large portion of the nation's forests were damaged or died out completely. In addition, Germany's geographic position has led to problems with transboundary pollution from neighboring countries.

Politically, Germany's system has unique features that have allowed activists access to policymaking on both formal and informal levels. Proportional representation has allowed the Green Party to gain seats in the two parliaments once they received a minimum of 5 percent of the national vote. German activists can also press their issues at the state level, with an opportunity to place issues of concern, such as water pollution, on the national agenda. Last, the German legal system is based on an independent judiciary that is considered wide open to interest groups. In recent years, the Green Party, die Grünen, formed in 1980, has lost much of its public support because the party's issues were absorbed by the mainstream parties. Most environmental organizations have switched from protests and demonstrations to more traditional tactics such as lobbying political leaders. The environmental community is concerned with a broad range of issues, mirroring the fact that the German public is considered one of the best informed on environmental topics in the world.

In France, the national green party, Les Verts, has maintained a consistent presence on the national level that rivals that of

Germany. The major national environmental organization, the Federation Nationale des Societies pour la Protection de la Nature, has piggybacked on antinuclear sentiment, whereas local groups appear to be focused on the border areas of the country. Their protests have centered around the issue of the import of hazardous waste—France is the world's largest legal net importer in the world. The government's regulation of environmental quality is considered weak by most observers, and private waste treatment companies are heavily subsidized by government agencies.

Latin America

Among the most publicized struggles involving environmental activists is the protection of the tropical rain forests and resources of the Amazon. Many in the industrialized world rallied behind Brazilian activist Chico Mendes, who was assassinated in 1988 (Mendes is profiled in Chapter 5). In a case similar to that of Mendes, another Brazilian found that placing one's life on the line to support indigenous people is one of the burdens of leadership. Activist Ademir Alfeu Federicci (Dema, as he was most commonly known) met a similar fate in 2001 when he was shot in the head in front of his wife and children in Altamira, Brazil. As coordinator of the Movement for the Development of the Transamazon and Xingu Region (MDTX), Dema worked to generate opposition to the construction of dams on the Xingu River and the illegal logging of the forests. He had previously worked in Catholic Church–based social movements and later became part of the agricultural union movement. MDTX and other organizations had denounced the corruption they had uncovered in the Amazon Development Department, alleging that project funds were being diverted to individuals. For his outspoken criticism, Dema was murdered, silencing many of those who had joined him.

This incident represents two important elements of environmental activism in Latin America. First, it is exemplary of many social movements that have had the support and backing of the Catholic Church, serving as an instrument of social change. The church's support (and sometimes its role as a sanctuary for activists) was a part of the reform movement in Latin America in the 1970s, and it continues today. It is not at all unusual for priests and other religious leaders to speak out against the government, sometimes at their own peril. Church leaders also assist in the process of organization building, providing training and financial assistance where possible. Second, many environmental activists participate at their own peril, and leaders are often silenced through violence.

In Ecuador, where activists battle not only deforestation but also a megaproject to build a 300-mile-long oil pipeline, strategies have focused on direct action. The project, Oleoducto de Crudos Pesados (OCP), is being directed by a multinational group of companies from Canada, the United States, Italy, Argentina, and Spain, with $900 million in loans from the German Westdeutsche Landesbank Bank. In July 2001, protesters rallied in front of the bank's offices in Germany and at the German Embassy in Quito. Realizing that they had little clout in attempting to influence the Ecuadorean government, which had already approved the project's environmental license, activists attempted to put pressure on the German state of North Rhine Westphalia, which holds a 43 percent interest in the pipeline. They drew support from Germany's well-organized Green Party, which subsequently lobbied the prime minister, Parliament, and the German minister of environment.

Activists have also engaged in direct action that mirrors the more radical groups in the United States, Canada, and Great Britain. Community residents, students, and NGO members focused on the Mindo Nambillo Cloudforest Reserve to prevent construction crews from entering the protected area. They held a peaceful encampment by occupying trees, chaining themselves to the base, or building platforms in the trees—the first such type of demonstration in Latin America. The protests were lauded by Julia Butterfly Hill, who occupied a platform in an old-growth redwood tree in northern California for two years.

The resistance movements in Quito and Mindo, organized primarily by Accion Ecologica and the Quito-based Oilwatch International, did not go unnoticed by the Ecuadorean government. A protest by women activists at the OCP offices became violent when company security guards assaulted the protesters and journalists covering the incident in order to prevent any favorable publicity. In February 2002, thousands of striking OCP construction workers and residents in northern Ecuador were attacked by government armed forces. The workers had been on strike to demand more funding for hospitals, roads, and clean running water. Erecting roadblocks, the workers had stopped the project despite a claim by the country's president that he would "bring war" to anyone in the way of the pipeline.

Dema's murder shows how violent the political mood has become when the issue is natural resources. Although some recognize hydroelectric power and timber industry activity as a sign of modernization, others fear that the cost of such development is too high, especially for local and native peoples. Killing the leaders of grassroots movements not only ends the influence of leaders but also stifles any further collaboration.

Peruvian farmers, who have few resources of their own to oppose projects they believe would negatively impact their communities, have relied upon global environmental organizations for assistance. In June 2002, the 36,000 residents of the Peruvian municipality of Tambogrande voted in a referendum (supported by Friends of the Earth International as election observers) to oppose a proposal to mine gold in their community. With 98 percent of the voters saying "no" to the proposal, the residents clearly made their views known by utilizing the referendum process that allowed them to vote on the proposal.

Prior to the referendum, resistance by citizen activists had been organized by a local leader, Godofredo Garcia Baca. In February 2001, Baca called upon local citizens to go on strike to mobilize in hopes of getting the Manhattan Mineral Corporation to withdraw its proposal. A month later, Baca was murdered under what have been called suspicious circumstances. The citizens' referendum served notice upon the company of their continued opposition to mining in their agriculturally based community.

Wildlife conservation served as the catalyst for environmental activism in Bolivia in the 1960s. Efforts were launched to protect the remaining populations of vicunas, a species related to the llama. A U.S.-trained researcher had monitored the rapid decline in the number of vicunas in the Bolivian Andes and sought to find out how other Latin American countries were protecting their wildlife. He developed a collaboration with a Chilean botanist, an Argentinean agriculture expert, and one of Peru's conservation pioneers to help draft an international agreement banning all trade in vicunas.

Bolivia's leadership in contemporary conservation activities can be tied to a number of factors that began in 1973 with the founding of the Bolivian Ecology Society. One leader used his political connections to propose a national park system—a project that could not have succeeded without the support of his friends and family members in the government. Many leaders received their initial training in university-level biological science programs, creating a community of concerned professionals with the expertise to identify potential environmental problems. Research led to activism—conferences held in the early 1980s led to the founding of the Environmental Defense League in 1985. Scientists became organizers, organizers formed coalitions, and coalitions recruited more members and volunteers. By the 1990s, Bolivia had gained prominence as a leader in biodiversity conservation.

In the Central American country of Costa Rica, the primary environmental issue is the preservation of biodiversity. On the one hand, Costa Rica has a long record of social activism and political participation. Its military was abolished in 1948, and the legislature is built upon proportional representation, giving more voice to minority interests. On the other hand, the people are sensitive to intervention by foreigners and a fear that their native culture will be dominated by outsiders.

As early as the mid-1800s, the government recognized the need to protect Costa Rica's biological wealth, with proclamations forbidding deforestation near watersheds, regulations on the use of wildlife, and proposals to protect marine life. The Garden Club, established to monitor the park created along the Pan American Highway, had a negligible impact on policy and was largely symbolic. But by the 1950s, university activists called upon citizens to conserve natural resources even at a time when the national government's policy was to boost agricultural production. Deforestation was considered "unpatriotic," with the land considered part of the people's national heritage. In 1950, the country celebrated Natural Resources Conservation Week, although the weak environmental movement had only minimal direct influence on government policies.

In the 1970s, activism and environmental policy reform were the norm, pairing biologists who monitored species loss with native Costa Ricans who sought to conserve and defend natural resources. Two men, Alvaro Ugalde and Mario Boza, were especially influential in establishing a national park system in the late 1960s after they visited the Great Smoky Mountains National Park in the United States. At the same time, conservation scientists, many of whom were inspired by visits to other countries where environmental movements were thriving, encouraged the Costa Rican tradition of civic organizing. A Costa Rican chapter of the Audubon Society was founded in 1971, followed in 1973 by Amigos de la Naturaleza and the Costa Rican Association for National Parks and Zoos.

One of the events that mobilized activists was a proposal to construct an intercontinental oil pipeline from the Caribbean to the Pacific, transporting oil from supertankers too large to navigate the Panama Canal. Likening the project to the Alaskan pipeline, activists warned of damage to coral reefs, the threat of oil spills, and security risks. Group members networked closely and lobbied the government to annul the project, which it did. Subsequent actions included student sit-ins to protest road construction, coalition building with teachers' union members, and seminars to the national organization of journalists on air pollution from a cement factory and deforestation near Arenal Volcano.

By the late 1980s, Costa Rican activists were well established, a situation that continues today. There are hundreds of organizations dealing with issues from urban water quality to campaigns to reduce the consumption of ozone-depleting chemicals. The media now sponsor special programs and broadcast public service announcements on environmental awareness, and environmental education has become a key part of the curriculum at both the high school and college level.

The environmental activists of the twenty-first century are similar in many ways to their U.S. counterparts. The new advocates are younger and did not participate in the national parks movement of the 1960s and 1970s. Many of them are attorneys, and the University of Costa Rica's environmental law emphasis prepares its graduates to participate in the Environment and Natural Resources Law Center. The Ambio Foundation conducts legal research, and the group Justice for Nature initiates litigation against polluters and the government to force compliance with environmental statutes. Other organizations, such as the Association for the Protection of Wild Flora and Fauna, conduct citizen patrols of national parks and mobilize volunteers who protest road building and logging. As a further sign of the growth of Costa Rica's environmental activism, there is now a national alliance of groups that networks and coordinates activities.

Mexico, Costa Rica's northern neighbor, has not experienced the same level of grassroots activism and environmental awareness. Rapid industrialization and a struggling economy have combined to create an environmental crisis that has taken a toll on the nation's natural resources. It is also a country where violence against environmental activists is common. Two leaders of an opposition movement against the Boise Cascade mill in Guerrero were arrested under what they believed were trumped-up charges in May 1999. The two men, leaders of the Peasant Ecologist Organization of Petalan and Coyuca de Catalan, complained that Mexican troops who stormed their village of Pizolta tortured them, charged them with drug and weapons possession, and linked them to the rebel group, the Popular Revolutionary Army. The men and their families believe they were arrested because they obstructed logging trucks. The company itself denied any involvement in the arrests; the mill was later closed for what Boise Cascade said were timber supply problems.

Despite President Vincente Fox's pledge to support environmental initiatives, the culture of Mexico does not encourage grassroots activism. Many of the nation's environmental organizations such as the Nature Conservancy, Ducks Unlimited, and Pronatura are considered elitist groups whose only real concern is habitat protection. The political system also provides only limited access for public participation, and the central government controls public forums that limit discussion. The greatest success appears to be along the U.S.-Mexican border, where local citizens can depend upon the support and resources of their U.S. counterparts. Issues such as the pollution caused by Mexico's *maquiladoras* and transboundary pollution generate much more media attention than problems within the interior. Still, most observers believe Mexico's environmental activism is still in its infancy, limited largely by the country's political and administrative systems.

Even though the activists in Latin America operate somewhat independently and recognize the risk that accompanies their participation, they are not totally isolated. The Sierra Club and Amnesty International have launched a campaign called Defending Those Who Give the Earth a Voice to monitor the plight of activists. The two leaders of the opposition to Boise Cascade were also supported by more than 100 activists from seven countries who sent a letter to the Mexican government demanding their release. Among those signing the letter were the American Lands Alliance, the Rainforest Action Network, and the International Forum on Globalization. By making such cases more visible, U.S. and global organizations attempt to make activism more transparent to the rest of the world and assist those who risk their life in environmental causes.

Oceania

Oceania comprises many nations in the South Pacific, and despite their geographic distance from activism in the United States and Europe, several countries have a long history of participation in environmental issues. Oceania is an example of a region that has also seen the involvement of many international organizations, described more fully below. Friends of the Earth–Australia (FOE-Australia), for instance, is a branch of the global group, but it focuses on issues particular to the continent. In recent years, FOE-Australia's members have been actively engaged in lobbying and advocacy, seeking World Heritage Site listing for the wetlands area near the Lake Eyre Basin, opposing uranium mining, and supporting indigenous communities. World Wildlife Fund–New Zealand has focused its conservation program on habitat protection, forest stewardship, and campaigns to save endangered species such as the North Island Hector's dolphin and the albatross.

Australia, New Zealand, and Tasmania were the birthplaces of the green party movement described in Chapter 1, although the emphasis on grassroots environmental activism continues to grow in other directions. The Australian constitution enacted in 1900 does not even mention the environment, and most regulatory programs were set up by the states, who used the U.S. Environmental Protection Agency as a model. Despite the decentralized nature of Australian government, most environmental protests have been directed at the commonwealth government rather than at the state level. In the early 1980s, thousands of activists were mobilized to oppose a proposal to build a huge hydroelectric dam in the Tasmanian wilderness, and eventually their opposition (and the support of the Australian High Court) was sufficient to convince the United Nations to declare the area a World Heritage Site.

More recently, the country's small Democratic Party has been able to take advantage of the compulsory voting requirement to attract environmental activists, and the proportional voting system used in the Senate has opened up slots for both the Democrats and the Green Party. In addition, activism is coordinated through the nation's largest organization, the Australian Conservation Foundation; farming groups; and branches of international groups such as Greenpeace. These groups participated in opposition to the siting of a waste incinerator at Corowa. Their protests, coupled with public opinion polls showing concerns about the transport of waste, ended the proposal.

In New Zealand, environmental activism is synonymous with the group Greenpeace, which has been involved in antinuclear protest in the area since the end of World War II. Greenpeace protesters have been roughed up by commandos, and their ship, *Rainbow Warrior*, was blown up by the French government in the Auckland harbor in 1985. But the maturity of the New Zealand environmental movement can be seen in the number of publications devoted to green issues, such as *New Zealand Environment, The Pacific Ecologist*, and the *Coconut Free Press*, which highlights environmental issues in the South Pacific. Aotearoa Independent Media Centre serves as an independent media outlet that enables groups throughout the region to distribute news quickly and effectively. These resources are vital to activists who are widely scattered throughout the area.

Another sign of New Zealand's support for activism is Environment and Conservation Organisations of New Zealand (ECO), a nonprofit network of sixty-five groups from around the country. ECO coordinates campaigns, serves as an information clearinghouse, and provides the public with environmental education programs. The New Zealand Trust of Conservation Volunteers serves an important role in the registration and coordination of volunteer conservation efforts. There are also activists involved in mainstream strategies as members of groups such as Save Animals From Exploitation (SAFE) and the Sustainable Cities Trust.

International Environmental Activism

In addition to the activism that has occurred at the local, regional, and national level, the past two decades have seen a tremendous burst of activity among transnational organizations. These groups, and their members, focus on issues that are often called common pool resources, such as the oceans and the atmosphere (global warming). The groups gained prominence at the 1992 United Nations Conference on Environment and Development in Rio de Janeiro, also known as the Earth Summit, where they held parallel meetings. Some of the activism has been linked to an increasing emphasis on a global economy and a global civil society that has broken down trade barriers and increased the spread of technology. Others feel that the increase in cooperative efforts stems from a growing concern about compassion for issues that transcend borders, such as population growth and poverty.

Equally important is the fact that U.S.-based groups have frequently provided the initial expertise, organizational resources, and seed moneys to develop counterpart or cooperative relationships with activists in other countries. For example, Earthjustice (which previously was known as the Sierra Club Legal Defense Fund) has formed a sister group in Canada, the Sierra Legal Defence Fund, and a partnership with Russia's Ecojuris. These organizations then become more globally focused, creating partnerships in several countries. In 1978, activists from Mexico, Brazil, Colombia, Venezuela, Bolivia, and Costa Rica toured the United States for one month to learn more about the strategies of U.S. environmental groups.

Western environmental groups have not been shy about their involvement in other countries. Typical of the intervention of nonnative activists is Kenya, where non-African researchers and administration specialists developed a plan for the preservation of the Amboseli basin at the base of Kilimanjaro. The local Maasai people had for decades killed wild animals that threatened their cattle as they grazed through the local vegetation. The government threatened the Maasai with various penalties for poaching wildlife on game reserves, and tourists were angry at everyone involved because they had come to see and photograph the wildlife and the Maasai. Social scientists and economists from North America, Europe, and East Africa gathered together to collaborate on a plan for integrated land use for the Amboseli Game Preserve. Their proposal to develop a small core park that would be turned over to the Maasai was met with criticism by most conservationists and the government, who preferred that the area be made a national park. Key to the success of the plan was the involvement of Maasai elders, the member of Parliament representing the area, local county councilors, and an American business leader who offered to donate funding for new wells outside the park where the cattle could be watered. Without their cooperation, the intervention by nonlocal people would have failed.

Just as there has been a growth in global environmental activism by NGOs, there has also been an increase in the number of trade associations and transnational corporations (TNCs) that participate in decision making, in the same way as their U.S. counterparts. One study estimates that there are now 38,000

TNCs with a total of over 250,000 affiliates worldwide (Clapp, 2001). These organizations and companies are giving more attention to the environment because they face increasing regulation and because they now operate on a global scale, requiring them to monitor environmental laws but also allowing them to trade more freely.

Typically, these economic/corporate organizations lobby their own governments when such issues as waste trade or global warming are being considered. They may participate in the deliberations over a proposed treaty, for example, to make sure any new regulations are favorable to their concerns and secure their interests. Some groups have been active in the development of voluntary environmental regulation, preferring to write and enforce their own practices and procedures rather than having them deliberated over by official decision makers.

Nowhere is this better demonstrated than in the ongoing debates over the transfer of hazardous wastes from developed to developing countries. TNCs were among the first to engage in toxic waste trading, both for recycling and for relocation to countries hard-pressed for cash and economic growth. Trade associations have sat alongside environmental organizations in the diplomatic talks that have dealt with waste trading, as well as participating in the more technical (and usually nonpublic) meetings that accompany them. The associations are influential in many respects, whether through providing expertise or providing side payments to gain favor with officials.

Many NGOs have invested heavily in conservation in developing countries. Organizations such as Conservation International, the Nature Conservancy, Rainforest International, and the World Wildlife Fund have purchased long-standing debts from creditor banks at a discount, using the money to purchase and preserve areas in debt-for-nature swaps. These programs have allowed global organizations to preserve and expand natural park areas, provide resources to limit poaching and illegal trade in endangered species, and train indigenous people in sustainable agriculture.

There has been considerable criticism, however, over the "meddling" of NGOs in the politics and culture of developing nations. Some observers believe that developing country governments are under pressure to accept hard currency at the expense of their sovereignty. The local population often has little influence or participation in decisions made by the central government, and intervention often diverts attention from the extraction of wealth and resources from the countries involved. Nonetheless, without the participation and resources of global organizations, many poorer countries would have little, if any, incentive to participate in environmental protection.

Sources and Further Reading

Adam, Hussein M. "Somalia: Environmental Degradation and Environmental Racism." In Laura Westra and Bill E. Lawson, eds., *Faces of Environmental Racism: Confronting Issues of Global Justice*, 2d ed., 203–227. Lanham, MD: Rowman and Littlefield, 2001.

"Antarctica." http://www.foei.org/antarctica. Accessed 4 June 2002.

Bomberg, Elizabeth E. *Green Parties and Politics in the European Union.* London: Routledge, 1998.

Bryant, Raymond L., and Sinead Bailey. *Third World Political Ecology.* London: Routledge, 1997.

Carter, Neil. *The Politics of the Environment: Ideas, Activism, Policy.* Cambridge: Cambridge University Press, 2001.

Chasek, Pamela S., ed. *The Global Environment in the Twenty-First Century: Prospects for International Cooperation.* New York: United Nations University Press, 2000.

Clapp, Jennifer. *Toxic Exports: The Transfer of Hazardous Wastes from Rich to Poor Countries.* Ithaca, NY: Cornell University Press, 2001.

Collinson, Helen, ed. *Green Guerillas: Environmental Conflicts and Initiatives in Latin America and the Caribbean.* Montreal: Black Rose Books, 1997.

Desai, Uday, ed. *Ecological Policy and Politics in Developing Countries.* Albany: State University of New York Press, 1998.

Diani, Mario. *Green Networks: A Structural Analysis of the Italian Environmental Movement.* Edinburgh: Edinburgh University Press, 1995.

Feshbach, Murray. *Ecological Disaster: Cleaning Up the Hidden Legacy of the Soviet Regime.* New York: Twentieth Century Fund, 1995.

"Green Belt Movement Founder on Environmental Activism." (February 2002 media release). http://www.mcgill.ca/releases/2002. Accessed 12 May 2002.

Hayes, Denis. "Editorial: Grassroots International Environmental Activism." *Electronic Green Journal* 15 (December 2001). http://egj.lib.uidaho.edu/egj15/hayes1.html. Accessed 21 March 2002.

Joyner, Christopher C. *Governing the Frozen Commons: The Antarctic Regime and Environmental Protection.* Columbia: University of South Carolina Press, 1998.

Miller, Marian A. L. *The Third World in Global Environmental Politics.* Boulder, CO: Lynne Rienner, 1995.

O'Neill, Kate. *Waste Trading among Rich Nations: Building a New Theory of Environmental Regulation.* Cambridge: MIT Press, 2000.

Paterson, Kent. "Mexican Environmentalists behind Bars." *Progressive* 64, no. 3 (March 2000): 21.

Pickerill, Jenny. "Environmental Internet Activism in Britain." *Peace Review* 13, no. 3 (2001): 365–370.

Poluso, Nancy Lee, and Michael Watts, eds. *Violent Environments.* Ithaca, NY: Cornell University Press, 2001.

Rock, Michael T. *Pollution Control in East Asia: Lessons from Newly Industrializing Economies.* Washington, DC: Resources for the Future Press, 2002.

Shull, Tad. *Redefining Red and Green: Ideology and Strategy in European Political Ecology.* Albany: State University of New York Press, 1999.

Steinberg, Paul F. *Environmental Leadership in Developing Countries: Transnational Relations and Biodiversity Policy in Costa Rica and Bolivia.* Cambridge: MIT Press, 2001.

Steiner, Andy. "Vandana Shiva: An Indian Physicist Who Fights for Small Farmers." *Utne Reader* (November–December 2001): 72.

Suliman, Mohamed, ed. *Ecology, Politics, and Violent Conflict.* London: Zed Books, 1998.

"Urge the Brazilian Government to Investigate the Recent Death of a Forest Activist and Ensure Protection for All Activists in Brazil." http://www.greenpeaceusa.org/save/alerts/dematext. Accessed 21 March 2002.

Wapner, Paul Kevin. *Environmental Activism and World Civic Politics.* Albany: State University of New York Press, 1996.

Weiss, Edith Brown, and Harold K. Jacobson, eds. *Engaging Countries: Strengthening Compliance with International Environmental Accords.* Cambridge: MIT Press, 2000.

Wells, Donald T. *Environmental Policy: A Global Perspective for the Twenty-First Century.* Upper Saddle River, NJ: Prentice-Hall, 1996.

Western, David. "In the Dust of Kilimanjaro." In David Rothenberg and Marta Ulvaeus, eds., *The World and the Wild: Expanding Wilderness Conservation beyond Its American Roots,* 65–79. Tucson: University of Arizona Press, 2001.

4

Chronology

Although interest in the natural world surfaced in many forms in the seventeenth century, most of those who have studied environmental problems and activism believe that activism is primarily a twentieth-century phenomenon. The century can also be divided into six distinct and significant periods, each marked by notable events and opportunities for interest groups to participate, as the chronology below indicates.

The first period is from 1900 to 1945, when specific interests began to mobilize against the environmental problems that came with industrialization. Nature and wildlife groups formed in both the United States and Great Britain, with smoke abatement groups established in major urban areas. Women played a strong part in the efforts to clean up cities as they took on the role of the "urban housekeeper." The major groups that were established during this time, such as the National Audubon Society, served as models for subsequent groups seeking to lobby, litigate, and educate. Other groups, such as the National Environmental Health Association, grew out of the growing professionalism of activists. There was, however, little interest in environmental activism in nonindustrialized countries until well after World War II.

From 1945 to 1960, leisure and recreational activities, along with rampant consumerism, placed additional pressure on wild areas and natural resources. This was the time when the first major environmental statutes were enacted and state-level environmental organizations began to form. Differences among the states led to variations in organizational clout and forms of activism. Those states with an existing and strong environmental culture, such as New York and New Jersey, brought together local and regional groups, making their efforts more focused and cohesive. National organizations such as the Sierra Club fostered additional coalition building by establishing chapters and by lending their financial resources and expertise to state and local groups. This resulted in a vertical structure from the local, to the regional, to the state, and to the national level. Scientific evidence related to the nature and causes of air and water pollution surfaced during this time, but most activists centered their attention on animals rather than air quality. During this second period, many professional and trade associations also began to consider the problems that were developing in the environment, although it was not until the next phase that they became activists.

The zenith of American environmentalism is believed to have begun in the 1960s, centered on 22 April 1970, the first Earth Day, and continued until around 1980. Through the administrations of Presidents Richard Nixon, Gerald Ford, and Jimmy Carter, there was a tremendous growth in the number of environmental organizations in the United States and fledgling activism in other countries. One of the reasons for that activism was the United Nations Conference on the Human Environment, which took place in Stockholm, Sweden, in 1972. Also included in this third period is the advent of the first green political parties throughout the world. More single-issue groups formed, especially those seeking protection for a single species, such as the International Crane Foundation and the Mountain Lion Preservation Foundation. Other groups, such as the Pesticide Action Network and Negative Population Growth, formed to deal with a specific problem.

President Ronald Reagan's administration marked a watershed in environmental policy and the resulting activism, which accompanied statutes enacted during that time that rolled back environmental protection legislation. From 1980 to 1988, Reagan almost single-handedly reignited the somewhat dormant environmental movement in the United States. There was a growth in grassroots citizen participation through such groups as the Citizens Clearinghouse for Hazardous Wastes. The groups alerted residents to chemical hazards in their community and developed a model for citizen monitoring of environmental problems. This fourth period showed little growth in new national organizations in the United States but an emerging environmental activism in developing countries and on the local level in the United States.

The fifth period, covering 1988 to 2000, was a time when environmental advocacy appeared to have peaked, and the movement lost momentum both in the United States and abroad. Although the Clinton administration did accomplish some progress from a policy standpoint (especially in his last year in office), there was no corresponding move forward in terms of activism. The exception would be 1994, when efforts by the Republican Party to limit environmental initiatives from the past created a hostile legislative arena. Many groups could not actively lobby Congress because of their status as charitable organizations under the Internal Revenue Service tax codes, and this reduced some of their effectiveness. There was, however, a growth in environmental organizations and activism on a global scale, with a growing awareness of worldwide issues. The period also marked the growth of groups dedicated to rolling back environmental protection legislation and regulations; such groups advocated wise use, county supremacy, and property rights.

Beginning with the election of George W. Bush in 2000, environmental policies began to shadow those of the Reagan administration. During his first few months in office, he made clear his intention to reduce the federal government's presence in environmental problem solving, expecting the states to take on a greater responsibility for management and funding. Initially, his advisers announced the possibility of circumventing or canceling environmental initiatives from the last year of the Clinton administration, arousing activists who felt that Bush's policies might look very much like those of Ronald Reagan. But on September 11, 2001, the world's attention shifted abruptly to a new item on the policy agenda—terrorism—which overshadowed virtually every other problem or issue. Billions of dollars began flowing to the defense of the nation and to impacted sectors of the economy such as the airlines. Environmental activists in the United States realized that for them, there would no longer be a podium for raising issues that just two years earlier had captured the attention of Congress and the nation.

The chronology that follows attempts to consolidate the period from 1900 to 2002, highlighting the major events that have occurred along with the activism that accompanied them. Although much of the chronology could just as easily have been benchmarked to the passage of major environmental statutes, such as the National Environmental Policy Act, legislation is more likely to be a product of activism rather than a catalyst for it.

1903   John Muir and Will Colby of the Sierra Club take President Theodore Roosevelt to Yosemite Valley and Mariposa Grove, California, lobbying him for the return of Yosemite to the national government. Muir becomes intrigued by the wilderness and beauty of the West and takes on responsibility for its protection.

1908   President Roosevelt holds the first White House Conservation Conference. A thousand participants, including governors of each state, meet to discuss the new concept of resource management. The conference turns out to be primarily symbolic, however. The National Conservation Commission, appointed by the president after the meeting, ends its work when Congress fails to appropriate funds for its operation.

1910   The Sierra Club lobbies for creation of Glacier National Park in Montana.

1911   Hundreds join the Women's Organization for Smoke Abatement in St. Louis, Missouri, to crusade against "the smoke nuisance" caused by smoking chimneys. Their actions become part of the efforts to clean up pollution caused by industrialization and growth in the country's urban areas.

1913   A private group, the National Committee for the Preservation of the Yosemite National Park, sends out protest literature to nearly 1,500 newspapers in opposition to the proposed Hetch Hetchy Dam across the Toulumne River in Yosemite. The dam would serve as a water source for the growing population of San Francisco. Many of the newspapers, especially in the West, publish editorials supporting wilderness preservation.

            The Sierra Club and other groups are unsuccessful in their efforts to stop the Hetch Hetchy project. The dispute pits John Muir against his friend, Gifford Pinchot, and Muir subsequently considers the failure one of his greatest disappointments and defeats.

1919   The National Parks Conservation Association is formed to support and monitor national parks on behalf of the public interest. The group replaces the National Conservation Commission formed by President Theodore Roosevelt.

1935   Aldo Leopold and Robert Marshall found the Wilderness Society. The group's goal is to work toward the long-term preservation of nature and wild resources. It is unique among other environmental organizations because it limits its political agenda to wilderness protection while other groups are becoming involved in a wide range of issues. The Wilderness Society adopts traditional political strategies such as lobbying members of Congress, mobilizing its members to participate in letter-writing campaigns, drafting legislation, and being involved in rule making.

1936   Sportsmen establish the National Wildlife Federation to protect land for fishing and hunting, adding a new dimension of environmental activism and support for wild areas. The initial focus on conservation education later broadens to include research and a raptor management program. It retains its primary identity as an educational group, supporting graduate fellowships and publishing material suitable for children.

1947   The Defenders of Wildlife is formed, initially to protect animal welfare and later to focus on endangered species and wildlife habitats. The group will eventually play a major role in the reintroduction of wolves in the United States and the restoration of wildlife refuges. It launches a three-V's campaign—Visit, Volunteer, and Voice concern for refuge protection—and becomes a supporter of a strengthened Endangered Species Act.

1948   William Vogt publishes *Road to Survival*, and Fairfield Osborne writes *Our Plundered Planet*; both are examinations of problems with overpopulation. They recommend an increased reliance upon birth control,

1948
(*cont.*)

prevention of waste, and finding new ways of stretching the food supply as the only way the human race will survive. Osborne follows up his ideas with another book, *The Limits of the Earth*, in 1953.

1949

Aldo Leopold's *Sand County Almanac* calls for an ethic to deal with man's relationship with the land, plants, and animals. The book becomes one of the major pieces of literature in the founding of the environmental movement, and Leopold becomes a public icon for his views. His approach brings a different perspective to the use of land, which, Leopold argues, does not belong to humanity.

1951

The Nature Conservancy is founded to purchase parcels of land, which might otherwise be developed or destroyed, in order to protect natural areas. The organization becomes a model for private land trusts throughout the United States.

Other activities include loans to local groups for land purchases, management of protected areas, and identification of potential parcels needing preservation. The organization purchases land in the United States, as well as land conservation projects in Canada, the Caribbean, and Latin America.

1956

Environmental groups successfully block construction of Echo Park Dam in Dinosaur National Monument. The plan is deleted from the Upper Colorado River Storage Project, an action that is considered one of the major victories of a coalition of organizations focused on a single issue.

1958

Timber workers in the Pacific Northwest oppose their employers' positions on wilderness preservation, one of the first efforts to mobilize trade groups on behalf of the environment.

1960

David Brower launches the Sierra Club publishing program that brings photographs of wilderness to the attention of millions of Americans, especially those unfamiliar with the West.

1961 The World Wildlife Fund is founded in Switzerland to preserve biological diversity and wildlife, becoming the world's largest international conservation organization. Its strategy is to use long-range action plans, to provide emergency assistance grants for rescue programs, and to direct scientifically based research on species such as the panda, peregrine falcon, Bengal tiger, and golden lion tamarin monkey.

Physicians for Social Responsibility is founded to eliminate nuclear weapons and the threat of nuclear war by providing public speakers, lobbying, and publishing research and analysis about nuclear war–related issues. The group shares the Nobel Peace Prize as a U.S. affiliate of Physicians for the Prevention of Nuclear War in 1985.

1962 Rachel Carson publishes *Silent Spring,* warning of the dangers of the use of the pesticide, DDT. Carson argues that human health may potentially suffer from the unknown effects of chemical insecticides. She is heavily criticized by the chemical and agriculture industries, who claim that she is not qualified to conduct research or that her findings are greatly flawed. The National Wildlife Federation, in contrast, awards her the title Conservationist of the Year in 1963. She dies in 1964 of breast cancer thought to be caused by chemical exposure and is posthumously given the Presidential Medal of Freedom in 1980.

Murray Bookchin writes *Our Synthetic Environment* to publicize how technology and pollution damage human health. He is known as an anarchist or libertarian for his controversial views and membership in the Communist Party in his early years. He criticizes groups such as Earth First! but remains a source of philosophical inspiration for radical environmental organizations.

1963   Biologist Barry Commoner's *Science and Survival* focuses attention on the dangers of radioactive fall-out from atomic testing. He founds the Center for the Biology of Natural Systems at Washington University to foster an interdisciplinary approach to studying government policy and environmental issues. His support for more government regulation results in his candidacy for president on the Citizens Party platform in 1980; he gathers nearly a quarter of a million votes but loses the election to Ronald Reagan.

1965   President Lyndon Johnson convenes the White House Conference on Natural Beauty, setting the stage for antilitter and roadside beautification efforts coordinated by his wife, Lady Bird Johnson.

1966   *Life* magazine publishes a photo article, "Concentration Camps for Dogs," after the president of the Animal Welfare Institute shows a photograph of an emaciated laboratory dog to the publisher. The article generates thousands of letters to the magazine and to Congress in support of the Laboratory Animal Welfare Act. The article reinvigorates the animal welfare activists in the United States and Great Britain.

1967   The Environmental Defense Fund is formed as a policy group, later becoming a key source of legal expertise, specializing in damage done by pesticides and advancing the concept of environmental rights. The group employs scientists and economists to develop new environmental policies on recycling, wetlands, toxic pollution, and rain forest protection. It becomes best known for its campaign to alert families to the dangers of lead poisoning.

The Internal Revenue Service rules that the Sierra Club will no longer have tax-exempt status because of its lobbying activities. The action is a direct result of advertisements Brower helped design and place in major newspapers to protest the building of dams. The loss of charitable status leads to a major division within the Sierra Club, forcing Brower out of the organization in 1969.

1968   René Dubois's *So Human an Animal* concludes that nature is being permanently damaged by humans and society. A French soil microbiologist, Dubois was also active in research to find a cure for tuberculosis and became interested in humans' ability to adapt. His theory about a healthy environment's being a natural right was coupled with a belief that people were not enemies of nature but had the potential to restore their environment.

The Sierra Club and other groups successfully block the building of a proposed dam and power-generating facility at Marble Canyon at the edge of the Grand Canyon.

The American Indian Movement is established to reclaim older treaty rights to land, water, fish, and wildlife resources taken by the government or private landholders. It will eventually serve as the catalyst for Native American activism to protect sacred lands and resources.

The Club of Rome, led by Italian entrepreneur Aurelia Peccei, brings together a small group of prominent industry leaders, who launch efforts to curb worldwide environmental degradation.

Paul Ehrlich's *The Population Bomb* reignites the controversy over population growth. He warns of the probability of world famine because humans are exceeding the earth's carrying capacity, calling for immediate action to control worldwide population as the only solution to the crisis.

1969   Santa Barbara, California, residents form Get Oil Out! (GOO!) in response to an offshore oil spill that damages marine life and beaches. The grassroots group demands that the offshore oil platforms be banned along the coastline to prevent future disasters.

1969
(cont.)

Former Sierra Club leader David Brower founds Friends of the Earth with an expanded environmental agenda. His controversial position on the siting of a nuclear power facility in California and maverick activities result in his ouster as executive director of the Sierra Club, a job he had held since 1952.

The Union of Concerned Scientists is formed, bringing science and technological issues into the nuclear power debate. The organization originates as a 1968 statement issued by faculty at the Massachusetts Institute of Technology that calls for scientists and engineers to assume responsibility for what their research produces.

1970

Teach-ins at the University of Michigan, modeled after antiwar events, are held in March. The kickoff event draws an estimated 15,000 students, corporate leaders, and numerous public officials, providing credibility for the issues raised by scientists and public health researchers.

Millions of people across the United States observe Earth Day on 22 April to focus public attention on environmental issues. Organized by former Stanford student Denis Hayes and with the political leadership of Wisconsin senator Gaylord Nelson, the event is modeled after the teach-ins of the Vietnam War. Earth Day becomes an annual event observed around the world.

Environmental Action is formed to lobby and alert voters to companies with bad pollution records. The group serves as one of the sponsors of Earth Day 1970 but later disbands.

The League of Conservation Voters is established to support pro-environment candidates by tracking congressional and executive branch actions and voting records. David Brower serves as an initial leader of the group and guides its formation. Its legislative scorecard remains one of the key elements of political campaigns and voter groups, and it becomes one of the organizations capable of making cash contributions to those it endorses. It also becomes a key information source for voters as it analyzes campaign advertising, looking for false or misleading statements, which it publicizes as "greenscam."

Attorneys Richard Ayres and John Adams found the Natural Resources Defense Council, providing technical, scientific, and legal assistance for litigation by grassroots groups. The organization becomes a key player in major federal legislation concerning air and water pollution, the protection of old-growth forests, energy, toxic chemicals, and habitat protection.

Activist and author Ralph Nader starts Public Interest Research Groups to study and recommend changes in public policy on behalf of consumers. The members of the groups are called Nader's Raiders for their enthusiasm and dedication to research and advocacy.

Massive flooding in the Alakananda Valley of India spurs the development of the Chipko Andalan, or "hug a tree" movement, led largely by women, to prevent logging and more flooding.

1971   The Don't Make a Wave Committee tries to sail to Alaska to protest nuclear weapons testing; the group later evolves into the international organization, Greenpeace.

The Sierra Club Legal Defense Fund is established to provide scientific research and legal expertise to force compliance with environmental laws and rules.

Successful demonstrations by local activists are held in Costa Rica to stop construction of a large dam and hydroelectric project in the Terraba Valley.

1972   Australian activists form United Tasmania Group, which attempts to block flooding of Lake Pedder and becomes the first Green Party.

1972
(cont.)

The Environmental Policy Center, a spin-off of the League of Conservation Voters, opens its offices in Washington, D.C., for advocacy and lobbying on water issues.

Stockholm's United Nations Conference on the Human Environment includes a nongovernmental organization conference to bring together international environmental leaders.

The U.S. Supreme Court decision in *Sierra Club v. Morton* allows environmental groups more power in filing lawsuits even when the group's members are not directly affected.

1973

An Arab oil embargo and resulting energy crisis focus attention on renewable and alternative sources of energy and support for groups advocating solar, wind, geothermal, and biomass energy. Millions of Americans line up for hours at gasoline service stations to fill up their cars with gas.

Labor unions stage the first environmental strike against Shell Oil.

The Ecology Party forms in Great Britain to make environmental issues the focus of an electoral strategy.

1974

An antinuclear conference spawns a network of organizations called Critical Mass to confront issues of nuclear power.

Economist Lester Brown founds the Worldwatch Institute as a research and publication center, focusing attention on the global nature of environmental issues. With a start-up grant of $500,000 from the Rockefeller Brothers Fund, Brown serves as executive director and produces the annual *State of the World* reports that provide current statistics and reliable information on global environmental issues.

1975   Greenpeace uses nonviolent confrontation to confront the Soviet whaling fleet off the California coast. The direct action organization becomes known for its efforts to block whaling vessels by positioning its members in small boats between whales and the hunters.

Environmentalists for Full Employment is formed to confront claims that environmental regulations would cost workers their jobs.

Wilderness lobbies and local citizens' groups force Congress to stop plans for Tocks Island Dam along the Delaware River between New Jersey and Pennsylvania that would have affected free-flowing rivers.

Widely published author Edward Abbey writes *The Monkey Wrench Gang*, which becomes the inspiration for radical environmentalism. The book describes the efforts of a small band of activists to damage equipment by pouring sugar and dirt into the gas tanks of vehicles at construction sites in the desert. Abbey's twenty-two books inspire Dave Foreman and other activists to coalesce into Earth First!

Paul Watson leads Greenpeace efforts to halt harvesting of baby harp seals off the coast of Newfoundland in a highly publicized action. His efforts help to focus international action on the highly emotional practice and bring more credibility to Greenpeace.

1976   The proposed Seabrook nuclear facility in New Hampshire leads to formation of an antinuclear power group, the Clamshell Alliance, and antinuclear demonstrations.

1977   Militant Greenpeace members leave to form the Sea Shepherd Conservation Society, led by Paul Watson. Watson and his followers believe that Greenpeace has compromised its goals and that direct action is the

1977 (cont.)	only way to make sure that laws protecting marine life are being enforced.

A Libertarian think tank, the Cato Institute, opens in San Francisco, providing commentary on social and environmental issues.

The Bombay (India) Environmental Action Group is formed to oppose construction of a fertilizer and petrochemical plant in a rural area. The group is made up of middle-class professionals.

1978 Residents of the New York subdivision, Love Canal, near Niagara Falls, begin to experience severe health problems, and toxic contamination is reported. Lois Gibbs and her neighbors find out that their 1950s-era homes were built on a toxic waste dump, and she forms the Love Canal Homeowners Association to seek information and compensation for her neighbors.

1979 The first Green Party candidate is elected to the national assembly in Switzerland.

Nevada state legislators enact the Sagebrush Rebellion Act, becoming leaders in opposition to federal wilderness policies.

Three Mile Island nuclear power plant reactor incident, near Harrisburg, Pennsylvania, provokes resurgence of antinuclear power sentiment.

1980 People for the Ethical Treatment of Animals is founded to oppose scientific research using animals and to oppose animal abuse and exploitation. The international nonprofit organization claims success in uncovering the abuse of animals in laboratory research and in the first conviction of an animal experimenter in the United States. Undercover videotape of laboratories, testing facilities, and exotic animal training schools sends a powerful visual image to the public.

Carolyn Merchant's book *The Death of Nature: Women, Ecology, and the Scientific Revolution* becomes classic reading in the development of ecofeminism. Her work as an environmental historian helps establish ecofeminism as a subject of credible debate, although there has been criticism of the concept as well. She attempts to bridge the gap between pragmatism and scholarly work.

Die Grünen party is formed by West German activists, eventually becoming one of the world's strongest green parties.

A Resources for the Future poll finds that 7 percent of Americans consider themselves as "environmentally active" and another 55 percent say they are sympathetic to the environmental movement.

The conservative Washington think tank, the Heritage Foundation, develops a blueprint to reduce government regulation of the environment—a plan later adopted by the Reagan administration.

Cosmetics industry leaders endow a $1 million research institute at Johns Hopkins University dedicated to finding methods to test products without using animals; the action is in response to nationwide demonstrations.

1981   The Citizen's Clearinghouse for Hazardous Waste is established by resident Lois Gibbs to organize local communities in response to pollution at New York's Love Canal.

Three 345,000-volt power poles are toppled near Moab, Utah, in an unsolved case of environmental sabotage.

Dave Foreman and other Earth First! members unfurl black plastic down the face of Glen Canyon Dam in Arizona to simulate a massive crack in the structure.

1981
(cont.)
The "cracking" event is widely publicized and provides Earth First! with the kind of media attention it relies upon for its efforts.

Two masked women calling themselves the People's Brigade for a Healthy Genetic Future burn a $180,000 helicopter in protest of aerial herbicide spraying in Oregon.

1982
Activists mobilize against hazardous waste in the first major national protest by black citizens in Warren County, North Carolina. White residents and black civil rights leaders come together to try to stop the building of a landfill near their community that would store PCB-contaminated soil. The effort does not succeed, but the protests become a model for other poor, minority communities seeking to bring attention to the issue of toxic racism and environmental justice.

The Pesticide Action Network forms as a coalition of 100 nongovernmental organizations throughout the world. Their goal is to persuade their governments to adopt stronger controls on pesticide use, teach consumers about pesticide hazards, and train farmers in alternative forms of production.

1983
The Urban Environment Conference in New Orleans brings together minority groups to discuss coalition building as a strategy.

The German Green Party passes the 5 percent vote threshold to gain seats in the Bundestag.

Environmental groups pressure the White House to force the resignations of EPA Administrator Anne Gorsuch and Interior Secretary James Watt. The two controversial Reagan appointees are accused of rolling back funding and enforcement of major environmental statutes in an effort to reduce the need for compliance by business groups. The protests are successful (both officials resign) and result in a membership boost for mainstream organizations.

The Asian People's Environment Network is established to collect and dispense environmental information among its 300-member nongovernmental organizations.

The Animal Liberation Front emerges as an ecoterrorist group, stealing dogs used in research on heart pacemakers from a medical laboratory in Torrance, California.

Samaj Parivartana Samudaya forms in India, organizing the rural poor for conservation and environmental restoration efforts, with some armed conflict between the group and the state government.

1984   Local residents organize to form the Hanford Education Action League to conduct research and public outreach regarding Hanford Nuclear Reservation in Washington. The group, called the Downwinders, seeks to obtain information on the health effects of the nuclear waste stored nearby. They track illnesses and birth defects in local residents to prove that the government knew about the potential for damage to human health.

Committees of Correspondence is organized in St. Paul, Minnesota, adopting ten key values of grassroots environmental groups.

A deadly leak at a Union Carbide facility in Bhopal, India, kills thousands, and groups ask for a review of toxic facilities in the United States.

Environmental groups are estimated to play a part in at least one-third of congressional elections; they are led by the Sierra Club and the Audubon Society, which computerize their membership lists by district to mobilize voters.

The Animal Legal Defense Fund is founded to use the courts to stop organized hunts, medical experimentation on animals, and classroom dissection projects.

1984
(cont.)

The Surfrider Foundation forms in California after surfers experience health problems later traced to violations of the federal Clean Water Act by paper companies.

Danube Circle is founded in Hungary to oppose the construction of a hydroelectric project on the Danube River.

The Competitive Enterprise Institute is founded to promote free market environmentalism designed to reframe the environmental debate.

1985

The French government sinks a Greenpeace vessel, *Rainbow Warrior*, in Auckland, New Zealand, harbor in protests over nuclear testing in the Pacific; later the government agrees to pay the group $8.5 million in damages.

The Earth Island Institute and other groups launch a campaign to stop all drift net and purse seine fishing by tuna fleets, threatening to boycott tuna products.

Greens achieve their first political success in New Haven, Connecticut, by receiving 10 percent of the vote, edging out Republicans as the city's number two party.

Friends of the Earth moves its headquarters from San Francisco to Washington, D.C., to have a stronger political presence in the capital. This action becomes a trend for major environmental organizations that become more professionalized, with leaders receiving large salaries and benefits. It will eventually result in membership backlash as grassroots organizers allege that the groups are out of touch with what is going on outside Washington, D.C.

Major environmental groups join forces to publish *An Environmental Agenda for the Future* with agreement on national environmental goals and policies.

Grassroots groups in California support passage of the antitoxics Proposition 65, the first successful environmental initiative since 1972.

1986   Explosion and fire at the Chernobyl nuclear facility in Ukraine send clouds of radiation over Europe and ignite antinuclear protests worldwide.

The Project for Ecological Recovery is founded in Thailand to oppose the government's policies that allowed and encouraged the exploitation of Thailand's natural resources.

1987   The United Church of Christ Commission on Racial Justice issues a report showing that minority communities bear a heavier burden for disposal of hazardous waste than the rest of the United States. The study reinforces the concept of environmental racism that is alleged to have occurred in the siting of unpopular facilities such as incinerators and landfills.

The McToxics campaign led by the Environmental Defense Fund against McDonald's Corporation pressures the company to eliminate its polystyrene clamshell food packaging. Other fast-food chains recognize the effects of potential boycotts and protests and begin to change their operations to make them more environmentally friendly.

Earth First! advocates ecotage—strategic, illegal resistance to "put a monkey wrench into the gears of the machine destroying natural diversity." The term stems from the book, *The Monkey Wrench Gang*, by Edward Abbey and becomes a code word for activists who damage property at the site of logging operations and construction in forests.

The first national conference of U.S. Greens in Amherst, Massachusetts, breaks up over deep ideological divisions.

The University of North Carolina becomes the base for the Student Environmental Action Coalition, reviving campus politics at universities and high schools.

1987
(cont.)
The Rainforest Action Network leads a boycott against Burger King restaurants to protest the use of beef raised in countries that destroy forestlands for grazing of cattle.

The Animal Liberation Front claims responsibility for setting a $3.5 million fire at a laboratory under construction at the University of California, Davis.

Two indigenous people's groups in Malaysia set up barricades across logging roads to prevent timber companies from logging on their traditional land and affecting their forest-dependent lifestyle.

Environmental Grantmakers Association is founded as an umbrella organization composed of over 200 corporations and foundations dedicated to environmental activism.

1989
The Environmental groups form Environmental Consortium of Minority Outreach to counter criticism that mainstream organizations are out of touch with minorities and their concerns.

Coalition for Environmentally Responsible Economies meets in New York to draft a ten-point environmental code of conduct for corporations, the Valdez Principles.

Massive oil spill in Alaska's Prince William Sound leads to a volunteer cleanup effort to save marine life.

1990
Redwood Summer demonstrations against logging in northern California are held despite counterdemonstrations by timber workers and agricultural interests.

Sea Shepherd Conservation Society targets drift net operations by Japan in the Pacific Ocean that kill marine life and birds.

Starkist and Chicken of the Sea tuna companies announce they will stop purchasing tuna caught by setting nets on dolphins or by using drift nets. Bumble Bee tuna follows suit a year later.

Two Earth First! activists are injured in the explosion of a pipe bomb in a car in Oakland, California.

Environmental groups mount an unsuccessful proposition, Big Green, in California elections. Industry and agricultural groups respond with the Big Brown initiative, outspending the environmental groups by nearly $13 million.

A Gallup public opinion poll finds that 76 percent of Americans consider themselves as environmentalists and half contribute to environmental organizations.

Earth Day 1990 marks the twentieth anniversary of the first Earth Day but is criticized as a media-driven event supported by corporate donors.

Alliance for America emerges as a coalition of over 100 groups and companies seeking to blunt the effectiveness of environmental activism.

TreePeople, a nonprofit group of urban foresters, plants more than 2 million trees in the Los Angeles area.

1991   Operation Bite Back launches a five-state arson campaign against the fur industry.

Five Earth First! defendants accused of plotting to sever power lines to nuclear plants in the West are sentenced in federal court.

1992   Parallel meetings are held in Rio de Janeiro, Brazil, to bring together an estimated 6,000 nongovernmental organizations during the United Nations Earth Summit.

The Sierra Club endorses Bill Clinton and Al Gore as Democratic Party nominees in the presidential election.

1992
(cont.)

A study estimates that there are 10,000 environmental groups in the United States, most of them locally based.

The Green Party gains ballot status in California elections.

League of Conservation Voters spends a record $600,000 in support of environmental candidates.

Students at Texas A & M University–Galveston successfully block the siting of a copper smelter on land owned by the University of Texas system.

1993

Environmental leaders criticize the North American Free Trade Agreement among the United States, Canada, and Mexico owing to fears of increased border pollution.

Earth First! sets up a campsite at Cove Mallard, Idaho, then leads protests against logging in the Nez Perce National Forest.

The Multinational People of Color Environmental Leadership Summit brings together grassroots groups to oppose toxic contamination and issues affecting minority communities.

In Tacoma, Washington's Pacific Lutheran University, Dirt People for the Earth leads students in a Green Games competition to reduce energy and water use in residence halls.

1994

The California Desert Coalition lobbies against the California Desert Protection Act as unnecessary locking up of wilderness areas.

USA Today reports that membership in the nation's top ten environmental groups has dropped dramatically from 1990 to 1993, with membership declining 44 percent for Greenpeace, 35 percent for the Wilderness Society, and 14 percent for the National Wildlife Federation.

1995   Property rights groups lead an unsuccessful fight in Congress seeking federal compensation for landowners affected by environmental regulations.

Environmental groups protest the salvage logging rider as part of Clinton administration timber policies.

Students at the University of Arizona gather to protest the building of an astronomy station in a remote area, chaining themselves to the gates.

In Florida, 100 middle school students discover that a baseball stadium is slated to be built in the habitat area of the endangered Florida panther. Through an organized effort, the group is able to convince the public to vote down the tourist tax that would have financed the stadium.

1996   Environmental opposition leaders oppose Clinton administration designations of new national monuments using the 1906 Antiquities Act.

Green Party USA nominates Ralph Nader for president at a convention in Los Angeles.

The Conservative Heartland Institute, based in Illinois, seeks out college activists to distribute free Earth Day newspapers "devoted to sound science and market-based environmentalism."

1997   Julia "Butterfly" Hill begins a two-year protest against logging in California's old-growth forests by occupying a redwood tree.

The Association of State Green Parties is established to support candidates for state-elected offices.

The Animal Liberation Front stages raids at fur farms, releasing thousands of mink into the wild.

1997 *(cont.)*	The Earth Liberation Front and Animal Liberation Front declare an official alliance, followed by at least twelve cases of arson and nearly $18 million in damage, mostly in the Pacific Northwest.
1998	The Earth Liberation Front claims credit for the $12 million arson of facilities at Vail Mountain, Colorado, to protest loss of habitat for wildlife.
1999	Students campaign at the University of Washington to urge the Board of Regents to divest any stock held in eight companies targeted as responsible for delaying progress to curb global warming.

Cascadia Forest Alliance members organize a blockade and tree sit demonstration urging the U.S. Forest Service to withdraw the sale of trees on the western slope of Mount Hood, Oregon.

The Wilderness Society publishes a list of fifteen most endangered wild areas in the United States to focus attention on deterioration of wildlife and their habitats.

The Animal Liberation Front sets fire at a meat company on Mother's Day to honor "Mother Earth and all the cows who have their babies stolen from them."

Alaska Youth for Environmental Action succeeds in persuading the Anchorage school district to cancel its annual spraying of schools with the insecticide carbaryl.

2000	The Anarchist Golfing Association claims responsibility for vandalism at a seed-testing facility in Oregon, retaliating against experiments with a genetically modified form of grass that could be used for putting greens on golf courses.

The Bluewater Network pressures the U.S. National Park Service to ban jet skis from parks, recreation areas, and seashores. The jet ski issue becomes a highly volatile crusade by owners who believe they should have the freedom to use their recreational vehicles on public lands and by environmental organizations and local residents who oppose both the noise caused in wilderness areas and the dangerous operation by some owners.

2001   Farmers in the Klamath Falls River Basin in southern Oregon sue the federal government to obtain water allocations for agricultural use. Government agencies claim the water must be diverted from farms to protect habitats of threatened fish. The diversion of water comes during a time of drought, forcing some farmers into bankruptcy because they cannot get water for their crops. Native American groups join the farmers to advocate water allocations for their use as part of a federal treaty.

Fruita, Colorado, enacts a three-mile open space buffer that extends around the town's periphery to slow down effects of growth. The action is exemplary of concern about urban sprawl and growth as the population, especially in the western United States, spreads out to allow for residential and commercial growth.

President George W. Bush revokes a Clinton administration proposal to reduce the allowable level of arsenic in drinking water. Major organizations begin to mobilize against the new administration's environmental policies and appointments. Subsequent studies by the National Academy of Sciences show that arsenic in drinking water is much more harmful to humans than originally thought. In October, the EPA agrees to accept the original Clinton plan.

The Earth Liberation Front claims credit for the arson of the University of Washington's Center for Urban Horticulture, causing $5.4 million in damage. States in the Pacific Northwest appear to be targeted for ecoterrorism, and newspapers provide more detailed coverage of the history of direct action in the region.

2001
(cont.)
Three environmental groups and the U.S. Fish and Wildlife Service agree to a temporary delay in critical habitat designations for eight species, which are frequently the subject of litigation, to allow the federal agency to free up funds for protection of twenty-nine other species.

Members of the Quechan Tribe in California's Imperial Valley file suit against an Interior Department decision to allow the development of an open-pit gold mine on land the tribe considers sacred.

2002
The U.S. Army Corps of Engineers revokes Clinton administration regulations on the restoration and creation of wetlands. Environmental groups criticize the Bush administration for ending the "no net loss" wetlands policy, whereas home builders and developers praise the action as a positive step toward ending unnecessary paperwork.

The Wilderness Society issues a State of the Environment report that accuses the Bush administration of "striking out on the environment."

California environmentalists protest a Bush administration decision to halt offshore drilling in Florida after refusing to do so for their state. Activists say the decision is a blatant attempt to boost the reelection efforts of George Bush's brother, Jeb Bush, for governor of Florida.

Apalachicola Bay and River Keeper file a lawsuit against the city of Apalachicola, Florida, for failure to prevent discharges from its wastewater treatment facility in violation of the Clean Water Act.

The World Wildlife Fund denounces Japan for its "shameful" participation in the fifty-fourth meeting of the International Whaling Commission, accusing the nation's officials of undermining whaling policies and preventing discussions of marine conservation.

Tens of thousands of participants attend the World Summit on Sustainable Development in Johannesburg, South Africa, in August–September. Parallel forums and events include nongovernmental organizations and protesters.

5

People and Organizations

As we saw in the first four chapters of this book, environmental activism involves a broad spectrum of individuals, groups, strategies, and issues. There is no such thing as an "average" activist, since there are differences in the amount of time available to participate, the expertise or interest level, the amount that can be contributed financially, and the goals the individuals or groups seek to pursue. This chapter outlines the efforts of twenty-five individuals and groups that range from the radical left, to moderate mainstream groups, to those termed *antienvironmental activists* who make up the environmental opposition. The selection process has created an overview of groups small to large, persons well known and others unknown, from the United States and other countries. Some have left a long-term legacy, whereas others' contributions have been a momentary catalyst for action. The goal is to show the diversity of their backgrounds, their successes and failures, and how they participated as activists, regardless of the problems or issues in which they became involved.

Melinda Ballard

She doesn't look like your stereotypical environmental activist. She drives a cream-colored Jaguar, something no self-respecting tree hugger could even consider. The Jaguar is filled not with shopping bags but with portable respirator masks. Her house in Dripping Springs, Texas, near Austin, might sound like a mansion, and in fact it is, with twenty-two rooms and over 11,000 square feet of living space on seventy-two acres of land. But Ballard, her husband, Ron Allison, and their son, Reese, are not living there anymore—and she has become a symbolic if reluctant warrior in the environmental health movement's battle against toxic mold contamination.

Mold? Yes, the icky, black or green slimy stuff that sometimes is found under beverage vending machines, in bathrooms, behind refrigerators, and when a roof leaks. There are thousands of types of mold, and their spores usually begin to grow when there is water leakage. Toxic molds such as Stachybotrys are the ones that scientists are most concerned about, because they secrete substances that can lead to a wide spectrum of health problems.

With a background in public relations and advertising in New York, Ballard became a mold expert by default. She had designed her house in Texas as a dream house. But when her family began experiencing a series of health problems—coughing up blood, memory loss, and her son's asthma—she realized that the family had visited doctors about fifty times during a three-month period. Her insurance company initially admitted there was a problem with mold and mildew that were trapped underneath the hardwood floors in the house, the result of a leak. She later discovered that the Texas governor's mansion was being remediated about the same time because the governor's wife, Laura Bush, had begun experiencing mold sensitivity. The cause was believed to be the Bush's air-conditioning system. In 1999, an experienced contractor who had a reputation for dealing with mold in commercial buildings advised Ballard and her family to leave their home as quickly as possible because tests had revealed pockets of toxic mold that could be related to cognitive impairment or other ailments.

Ballard's claims against insurance companies began to build, and she and her husband originally filed for $100 million in damages. In June 2001, a jury agreed to an award of $32 million in a lawsuit against the company that had insured her home, a figure that included $12 million in punitive damages and $5 million for mental anguish. Her lawsuit alleged that the insurance company had mishandled her claim. A spokesperson for the firm credits Ballard with having started a heightened interest in mold, and Ballard is already spending some of the funds she expects to receive to gather together scientists, doctors, and others who study people who suffer from mold poisoning. Environmental activists note, however, that Ballard could not have fought the fungus battle without her personal fortune and resources.

One of those people who heightened interest in mold as an environmental health problem is Erin Brockovich, the subject of a Julia Roberts film who has begun an environmental crusade of her own. Brockovich has learned that her twelve-year-old Los Angeles–area home is also filled with mold. Brockovich says she has suffered from chronic headaches, facial rashes, and other problems that have caused her to miss speaking engagements and other work for more than a year. In March 2001, she testified before a California legislative committee about her experiences, which she believes are due to water damage in the home.

There have been a number of consequences since Ballard's situation was publicized in the national media. Thousands of other home owners have filed mold-related claims with their insurance companies, school officials throughout the United States have closed down entire schools and buildings, and thousands more lawsuits are now pending. A group called Homeowners Against Deficient Dwellings (HADD) is working to help other home owners who have had experiences similar to those of Ballard and Brockovich. Ballard's experience as the unofficial leader of what some have called a new grassroots consumer movement has ignited her activism—she decided to campaign for a seat in the Texas legislature so she could sit on the committee that oversees the state's insurance companies.

Sources and Further Reading

Belkin, Lisa. "Haunted by Mold." *New York Times Magazine,* 12 August 2001.

Farley, Rose. "Blame Games." *Dallas Observer Online.* http://www.dallasobserver.com. Accessed 13 September 2001.

LePage, Andrew. "Activist Explains Dangers of Toxic Mold." *Sacramento Bee,* 8 March 2001.

Mullen, Francis X., Jr. "Toxic Mold: Lawyer Compiles Growing Case File." *Reno Gazette-Journal,* 14 January 2001.

Judith Beatrice Bari (1949–1997)

"Don't Mourn, Organize!" was one of the slogans Judi Bari wore on a T-shirt as an environmental activist. Whether she best fits the description of a radical or an ecoterrorist, she undeniably had a major impact on the protests against the logging of old-growth redwood trees in California. She died in 1997 at the age of forty-seven of breast cancer, having lived a life that fellow activist

David Brower called ending much too soon. She had refused chemotherapy and hospitalization for the cancer, telling friends that she wanted to die at home in her mountain cabin near Willits, in northern California. She had asked friends to list her occupation as "revolutionary" in her obituary.

Bari's activism began in the late 1960s while she was a student at the University of Maryland. She, like many other college students, participated in protests against the war in Vietnam and dropped out of the university in 1972. She moved on to labor union organizing, using as a base her job at a large grocery chain and becoming the union shop steward there. This was followed by a job change, working at the U.S. Postal Service Bulk Mailing facility near Washington, D.C. She continued her union activities, organizing a wildcat strike by workers seeking better working conditions.

She moved to California in 1979, was married, and lived north of San Francisco, where she became active in protests against U.S. intervention in Central America and in antinuclear groups. She and her husband divorced, and she lived in Mendocino with her two daughters, Lisa and Jessica. While working as a carpenter, she learned that the redwood boards being used at a house she was building had come from old-growth redwood trees, some of which may have been several thousand years old. She turned her attention to preserving the great redwood forests of northern California and by 1988 became a key leader and campfire fiddler in the new collective of activists called Earth First!

She was an extremely successful community organizer, using the strategies she had learned in her work with labor unions to unite thousands of demonstrators. She teamed up with fellow Mendocino activist Darryl Cherney to make Earth First! more diverse, clinging to the concept of deep ecology and also becoming involved with abortion rights demonstrations. She also had a background in music and graphic arts and used those skills to rally protesters with songs she had written.

Bari is best known, though, for two specific incidents. The first took place in 1990, when she and Cherney developed a plan to bring thousands of college students from all over the country to northern California for the Redwood Summer project. They modeled the event on similar actions that had been accomplished in the South during the civil rights movement. Their goal was to help support a statewide voter initiative called Forests

Forever. The plan attracted the attention of timber companies, who were alleged to have engaged in sending death threats and convincing a truck driver to intentionally damage her car and run it off the road. The police seemed to pay little attention to her complaints, which Bari believed were part of an effort to discredit Earth First!

The second incident took place shortly thereafter. Bari and Cherney were driving her car in Oakland, California, when a bomb exploded under her seat, fracturing her pelvis and causing major nerve and tissue injuries from which she would never fully recover. Within hours, the two activists were placed under arrest and were being investigated by the Federal Bureau of Investigation (FBI), with bail set at $100,000 each. They were accused of knowingly transporting a bomb as part of their ecoterrorist activities and were later accused of having made the bomb themselves, based on evidence found in Bari's house and car. Despite the law enforcement agencies' claims, however, neither Bari nor Cherney was ever charged with a crime, and no other suspects were ever held or arrested.

Because they believed they had been unjustly arrested, the two filed a civil rights suit against the FBI, the Oakland Police Department, and several individuals. The suit claimed that the bombing had been part of a conspiracy to silence Bari and that the agencies had refused to believe or investigate Bari's claims about the death threats. It also alleged that their constitutional rights had been violated and denied equal protection. After Bari's death in 1997, the suit progressed through the federal court system, and the couple's attorney even asked the attorney general, Janet Reno, to appoint a special prosecutor to handle the case. The Redwood Justice Fund originally founded by Judi Bari continues its work in Santa Rosa, California. In June 2002, a federal jury awarded $4.4 million to Cherney and to Bari's estate.

Sources and Further Reading

Bari, Judi. *Timber Wars*. Monroe, ME: Common Courage Press, 1994.

Bernstein, Dennis. "On Guard: Remembrance, Judi Bari." *San Francisco Bay Guardian*, 5 March 1997.

"Redwood Summer Justice Project." http://www.judibari.org. Accessed 18 September 2001.

Zakin, Susan. *Coyotes and Town Dogs: Earth First! and the Environmental Movement*. New York: Penguin USA, 1995.

Blue Ribbon Coalition

Environmental activism can take many forms. One of the organizations that believes that its activities are designed to protect precious resources in the same way as other groups is the Blue Ribbon Coalition (BRC). Its name tells little about its mission, but its motto, Preserving Our Natural Resources FOR the Public Instead of FROM the Public, explains a different perspective on mainstream environmental groups' activities. Its values are, it believes, the same as those of other environmental organizations, including tolerance, land stewardship, equity and fairness, and education.

The membership of the BRC truly represents a diverse constituency. One element is made up of statewide off-highway vehicle users, such as the Alaska State Snowmobile Association, the Arizona Association of 4WD Clubs, and the Georgia Recreational Trail Riders Association. Many of the groups in the coalition represent local recreation interests, such as the Cadillac, Michigan Winter Warriors, the McCall Area Snowmobilers in Idaho, and the Acadiana Dirt Bike Riders of Louisiana. Small businesses catering to outdoor recreationists have joined (D and H Cycle, Cullman, Alabama; Bohannon Auto Service, Bentonville, Arkansas), as have companies that appear to have no direct interest, such as Big K Concrete Cutting in Sunland, California; Arizona Interior Textures in Phoenix; and Advanced Stair and Rail in Orlando, Florida. Individuals can also join; among those who have are a chiropractor, a dentist, and attorneys.

Some BRC interests sound as if they are more traditional environmental organizations, although their interests seldom converge. The Mineral King Preservation Society, Save Orrick Committee, and Shawnee Trail Conservancy sound more as if they would restrict access to public lands rather than being part of a national organization devoted to increasing recreational access.

But the most controversial membership element is represented by larger corporate interests, who some argue are the real power behind the BRC. They represent the major names in recreational vehicles (American Suzuki Motor Corporation, Kawasaki Motor Corporation), resource extraction companies with an interest in opening up wilderness areas (Sierra Forest Products, Associated Logging Contractors, Boise Cascade Corporation), and grazing interests (J. R. Simplot, Joyce Livestock Company). The United States Public Interest Research Group accuses the companies funding the BRC of intensive lobbying efforts, including more than $46 million (1997–1999), with an average of 146 lobbyists working on their issues at any time during the period (U.S. Public Interest Research Group, 2001). They also make major contributions to political candidates, primarily to current members of Congress.

Like other interests, the BRC has a presence in Washington, D.C., that focuses on the legislative and judicial system. It takes credit for the passage of the Symms National Recreational Trails Act that sets aside a portion of federal gas tax revenues for multiple-use trails and facilities throughout the United States. The BRC's Legal Action Fund supports dozens of litigants seeking to open up public lands for additional use, and its magazine reaches a wide audience with updates on legislation, court cases, and public hearings.

In some ways, the group's goals are not unlike those of other organizations that deal with wilderness issues, as well as ethical concerns. The BRC condemns the illegal use of alcohol and drugs while driving—an objective it shares with Mothers Against Drunk Driving (MADD). The BRC supports safety education programs for youth and adults, which is common to most hunting organizations. Somewhat ironically, it agrees with many environmental organizations that have opposed fees for access to public land. Those same groups often engage in one of the BRC's major family activities: trail reconstruction and cleanup.

One of the primary areas where the Blue Ribbon Coalition's members are at odds with other groups is over the designation of national monuments that their leaders believe are "land lock-up schemes" ("Environmental Groups Fight," 2001). During the late 1990s, President Bill Clinton used his powers under the 1906 Antiquities Act to create national monuments, primarily in the West. The BRC filed suit in an attempt to invalidate the president's proclamations of several of the most controversial designations, including the Canyon of the Ancients in Colorado and the Ironwood Forest in Arizona. Several environmental organizations, including Earthjustice, the Natural Resources Defense Council, the Wilderness Society, and the National Wildlife Fund, filed requests with the District Court of Washington, D.C., to intervene. A similar suit against the designation of the Giant Sequoia National Monument was dismissed by the court in 2001, holding that there was no evidence the president had violated the law when the 327,000-acre monument was established.

The BRC is well funded and well organized and is one of the most visible of the groups at the conservative front of environmental activism. Through its ability to quickly mobilize its coalition members, it represents an unusually powerful voice in the legislative and judicial arenas.

Sources and Further Reading

Deal, Carl. *The Greenpeace Guide to Anti-Environmental Organizations.* Berkeley, CA: Odonian Press, 1993.

"Environmental Groups Fight to Prevent Attack by the Blue Ribbon Coalition on the Antiquities Act." http://www.ecoworld.com. Accessed 5 October 2001.

Helvarg, David. *The War against the Greens: The "Wise Use" Movement, the New Right, and Anti-Environmental Violence.* San Francisco: Sierra Club Books, 1994.

Share Trails. http://www.sharetrails.org.

U.S. Public Interest Research Group. "Blue Ribbon Coalition." http://www.pirg.org/reports/enviro. Accessed 5 October 2001.

Norman E. Borlaug (1914–)

In recent years, the Nobel Peace Prize, one of the world's most prestigious awards, has gone to activists, leaders, and organizations that have worked to bring peace to their region or world. But in 1970—the same year that the first Earth Day was observed—the prize was awarded to a U.S. scientist who has been a central figure in the green, or agricultural, revolution.

The Iowa-born researcher studied in a variety of academic disciplines. Norman Borlaug earned a bachelor's degree in forestry from the University of Minnesota in 1937, then worked for the U.S. Forest Service before returning to the university to receive a master's degree and doctorate in plant pathology in 1942. Initially, he worked as a microbiologist for the Du Pont de Nemours Foundation, supervising research on industrial and agricultural compounds used to eliminate fungi and bacteria.

His work changed dramatically when he accepted an opportunity to work with the Cooperative Wheat Research and Production Program in Mexico, a project jointly sponsored by the Rockefeller Foundation and the Mexican government. Along with other scientists representing fields as diverse as genetics and cereal technology, Borlaug assisted in the development of wheat that was disease resistant and produced a higher yield.

The project that made Borlaug noteworthy, however, was his diligence in making new cereal grains available to developing countries suffering from famine and drought. Focusing on building successful crop acreage in Mexico, Pakistan, and India, he helped to transfer agricultural techniques to other nations in Africa, the Middle East, and Latin America. His work was supported further by grants from both the Rockefeller and Ford Foundations, the establishment of the International Maize and Wheat Movement Center, and his appointment as director. The international research and training institute brought together young scientists from all over the world who worked to increase harvest production. A second major project was Borlaug's attempts to produce a type of grain that would be even more nutritious than wheat and that would be higher yielding. This involved research on a human-produced grain, triticale.

Although Borlaug has received many honors throughout his career, from honorary university doctorates to the Nobel Peace Prize, he realizes that his efforts cannot solve the more serious problems of population growth and the need for better systems of food distribution. In more recent years, he has been criticized by scientists and policymakers who believe his research may have unintended consequences because of the manipulation of native seeds and the use of additional agricultural chemicals. Still, Borlaug is considered a practical humanitarian who admits he would rather work in the soil than chase what he calls "academic butterflies" ("Norman Earnest Borlaug—Biography," 2001).

Sources and Further Reading

Bickel, Lennard. *Facing Starvation: Norman Borlaug and the Fight against Hunger.* New York: Dutton, 1994.

Brown, Lester. *Seeds of Change: The Green Revolution and Development in the 1970s.* New York: Praeger, 1970.

Freeman, Orville. *World without Hunger.* New York: Praeger, 1968.

Hardin, Clifford M., ed. *Overcoming World Hunger.* Englewood Cliffs, NJ: Prentice-Hall, 1969.

"Norman Earnest Borlaug—Biography." http://www.nobel.se/peace/laureates/1970/borlaug-bio.html. Accessed 28 September 2001.

Wharton, Clifton R., Jr. "The Green Revolution: Cornucopia or Pandora's Box?" *Foreign Affairs* 47 (April 1969): 464–476.

David Brower (1912–2000)

David Brower's biographer, John McPhee, called Brower the archdruid in the 1971 book, *Encounters with the Archdruid,* an affectionate title that would accompany Brower throughout his life. When Brower died at age eighty-eight, he left behind a legacy of environmental activism unmatched in the twentieth century. As a conservationist, leader, defender of wildlife, Nobel Peace Prize nominee, and tireless advocate for the environment, he became an icon and role model in the environmental movement.

Born in Berkeley, California, Brower was introduced by his family to hiking and nature as a child. He began his college studies at the University of California, Berkeley, in 1929 but dropped out without finishing his degree. His antipathy toward academia was lifelong, and he frequently railed against scholars who spent too much time in their offices instead of in the field. Brower was first known as a mountaineer who developed routes across many of the nation's prime climbing areas in the 1930s. His interest in mountaineering drew him to the Sierra Club, which at the time was primarily a hiking club without a political agenda. He wrote articles for the group's newsletter, campaigned for the inclusion of Kings Canyon as a national park, and in 1943 married his wife, Anne, who worked with the University of California Press. During World War II, Brower's service commitment was to teach U.S. troops his mountaineering skills.

When he returned to the United States after the war, he was named the executive director of the Sierra Club in 1952. He had earned the job by advocating additional protection of Kings Canyon by limiting road building, an issue that had split the organization. At the time, the organization was primarily based in California, with an older and more conservative board of directors. Brower rapidly crafted a new agenda, and the Sierra Club found itself at the forefront of desert protection and an opponent of dam building. One of his first major victories was a fight to block the building of a proposed dam in Dinosaur National Monument. He has said that his life's major disappointment was a compromise agreement he made with the government to allow a dam to be built in Glen Canyon in exchange for the demise of the dam proposal in Dinosaur National Monument.

Just as the environmental movement was gaining momentum in the 1960s, Brower and the Sierra Club popularized the protection of wild places through a nineteen-volume series of coffee-table books featuring the work of the nation's top photographers. Millions of Americans became familiar with environmental issues by seeing the work of Ansel Adams, Eliot Porter, and others who brought the images of Yosemite, the Sierra Nevada, and Southwest deserts into the homes of previously uninterested middle- and upper-class families.

A tireless crusader, Brower's advocacy agenda moved quickly from organizing hiking trips to using the tools of public relations and lobbying. Working in conjunction with the Wilderness Society, the Sierra Club successfully sought passage of the Wilderness Act of 1964, legislation that would pave the way for the designation of millions of acres of land as permanent wilderness. The speed at which he sought change proved to be too rapid for many of the members of the Sierra Club board of directors. The final straw in destroying his career as president came after Brower single-handedly made the decision to place highly politicized full-page newspaper ads he had conceived with an advertising agency, causing the Sierra Club to lose its coveted nonprofit status. He was fired in 1969 as a result of his actions.

Undaunted, he founded a new group, Friends of the Earth, in 1969, expanding the environmental agenda even more by including international issues. Friends of the Earth led to Brower's involvement with the League of Conservation Voters, another vehicle for publicizing environmental issues and the voting records of members of Congress. Once again, his vision was not shared by the group's leadership, and he was fired. Moving on, he served as chair of the board of directors of a new group he founded, Earth Island Institute, where he worked until his death in his hometown of Berkeley.

In the last ten years of his life, Brower hit the road as a highly sought-after speaker, traveling from small Oregon colleges to international conferences. His major theme was the need for what he called environmental CPR: Conservation, Preservation, and Restoration. He was especially interested in speaking to college audiences, which he believed would lead the next generation of environmental advocates. He and Anne almost always traveled together and sometimes spent the time between speeches walking and hiking in the local areas where they stayed. Hand in hand, they pondered the next big battle.

Sources and Further Reading

Brower, David. *For Earth's Sake: The Life and Times of David Brower.* Salt Lake City: Peregrine Smith Books, 1990.

Cohen, Michael P. *The History of the Sierra Club, 1892–1970.* San Francisco: Sierra Club Books, 1988.

Fox, Stephen. *John Muir and His Legacy: The American Conservation Movement.* Madison: University of Wisconsin Press, 1985.

McKibben, Bill. "Remembering the Twentieth Century's Greatest Environmentalist." *Rolling Stone* 28 (December 2000): 33, 36.

Tuner, Tom. *Friends of the Earth: The First Sixteen Years.* San Francisco: Earth Island Institute, 1986.

Helen Broinowski Caldicott (1938–)

One of the professional groups that is sometimes overlooked as environmental activists is doctors. Helen Caldicott, a physician who originally specialized in cystic fibrosis at Harvard Medical School in Boston, is an exception. She is noted not only for her individual efforts on behalf of the environment and against nuclear dangers but also for founding Physicians for Social Responsibility. The organization, which has counterparts in other countries, is linked to International Physicians for the Prevention of Nuclear War, which was presented the Nobel Peace Prize in 1985.

Born in Melbourne, Australia, she entered the University of Adelaide Medical School in 1956, where she was awarded her medical degree; she practiced medicine in Adelaide. When she and her husband, William, moved to the United States in 1977, Caldicott served on the staff of the Children's Hospital Medical Center in Boston. She worked there for three years until she resigned her position to devote her efforts full time to the prevention of nuclear war. She cites as her inspiration for activism Neville Shute's novel, *On the Beach,* which deals with an accidental nuclear holocaust, and the story of Robin Hood, who takes from the rich to give to the poor.

During the 1970s and 1980s, she led several successful environmental campaigns, working to ban atmospheric nuclear testing in the South Pacific, stopping the mining of uranium in Australia, and forcing the government there not to sell or transport uranium from 1975 to 1982. When she perceived that Physicians for Social Responsibility was too conservative for her political views, she left the group. Returning to Australia in 1986, she founded a new political party, Our Common Future, and in 1987 ran unsuccessfully as an independent for the federal parliament. She also founded the group Women's Action for Nuclear Disarmament in the United States in 1980.

Although she has received numerous awards and honorary degrees for her work in advocating for changes in human behavior to stop environmental destruction, she admits that the accomplishments have carried with them a personal cost. In her 1996 autobiography, *A Desperate Passion,* she writes that her activism has been difficult for her three children, her medical career, and her personal well-being. She has been criticized by the Environmental Policy Task Force of the National Center for Public Policy Research, which characterized her positions and rhetoric as "irresponsible and extreme" ("Environmental Scientist," 2001). She admits that "peace and survival are my passions" ("Helen Caldicott, M.D.," 2001), continuing her political involvement, writing, and lecture tours while splitting her time between Australia and the United States.

Sources and Further Reading

Caldicott, Helen. *Missile Envy.* New York: Morrow, 1984.

———. *Nuclear Madness.* Rev. ed. New York: W. W. Norton, 1994.

———. *A Desperate Passion: An Autobiography.* New York: W. W. Norton, 1996.

"Environmental Scientist: Dr. Helen Caldicott." National Center for Public Policy Research. http://www.nationalcenter.org/dos. Accessed 28 September 2001.

"Helen Caldicott, M.D." http://www.noradiation.org/caldicott. Accessed 28 September 2001.

Co-Op America

Even before Earth Day 1970, many environmental groups were encouraging their members to shop responsibly by making choices based on sustainability, resources, and the reduction of consumption. Co-Op America (CA), a nonprofit organization founded in 1982, has a long history of informing consumers about the social and environmental records of companies that are stewards of the environment and promote healthy communities. Based in Washington, D.C., CA has an estimated 50,000 individual members and more than 2,000 business members ("Co-Op America," 2001).

The organization is focused on four programs: Green Business, Consumer Education and Empowerment, Corporate Responsibility, and Sustainable Living. It also publishes the *National Green Pages*, a yearly directory of environmentally responsible businesses. The publication also tells members of the public how to green their homes and how to use green spending strategies and socially responsible investing. Other publications report ongoing and new boycotts, give tips on how to be an environmentally conscious shopper, and advise business members how to respond in a competitive marketplace. The group is especially interested in forest preservation, distributing a free WoodWise consumer guide that provides practical solutions such as buying tree-free paper, stopping junk mail, and not purchasing teak furniture. Over 200,000 copies of the guide were initially distributed in 2000, and celebrities such as Chevy Chase have lent their support to the effort, which involves sixteen leading environmental organizations.

The CA Business Network, although not unique, is an important aspect of the group's activism. Members of the network are evaluated by the organization to determine the company's commitments to four primary stakeholders: customers, employees, community, and the planet. The network provides tools that businesses can use to expand their market and reduce the costs of conducting business in a responsible way. For example, suggestions for creating a Green Office highlight energy-saving office equipment; natural lighting; and nontoxic, recycled carpets, wall covering, and furnishings. This is stressed as an important move toward providing employees with a safe and comfortable working environment, which can increase productivity and reduce absenteeism. The group also urges consumers to support businesses that create jobs rather than those that exploit workers and to purchase fair trade products.

For consumers, CA serves as a gateway between businesses and the public. In 1998, the group encouraged consumers to write letters and contact Home Depot stores urging the company to start carrying certified, sustainably harvested lumber. They also support boycotts of companies believed to conduct business in an environmentally irresponsible manner. The boycotts are seen as an effective strategy for enacting social change. Although not directly endorsing specific boycotts, CA tracks them and lists them on their web site and newsletter, *Boycott Action News*. The group also works closely with other environmental organizations as part of a coalition to protect forests worldwide. Partners include the Certified Forest Products Council, Conservatree, Fiber Futures, the Independent Press Association, Rainforest Action Network, and ReThink Paper, a project of the Earth Island Institute. As one CA official puts it, "Our mission is to create a more just and sustainable society by using economic power" (Macauley, 1998).

Sources and Further Reading

Coolidge, Shelley Donald. "Boycotts of Companies Grow as a Protest Tool." *Christian Science Monitor* 88, no. 246 (15 November 1996): 1.

"Co-Op America: Who We Are and What We Do." http://www.coopamerica.org. Accessed 13 November 2001.

Kaye, Steven D. "Attention, Investors—Agenda Ahead." *U.S. News and World Report* 120, no. 17 (29 April 1996): 83.

Macaulay, Catherine. "Buying Green." *E Magazine* 9, no. 5 (September–October 1998): 18–23.

Council for Responsible Genetics

With new developments in cloning, issues regarding genetic technologies are coming to the forefront for activists who are concerned that introduced engineered organisms could alter natural systems, affect the environment in negative ways, and do all of this without much public knowledge or interaction. One concerned group that tackles such issues is the Council for Responsible Genetics (CRG). The council is a nonprofit, nongovernmental organization founded in Cambridge, Massachusetts, in 1983 and is made up of scientists, environmentalists, public health advocates, physicians, lawyers, and other concerned citizens.

The CRG monitors the development of new genetic technologies and their application through two main program areas: human genetics and commercial biotechnology. In tracking current genetic technologies, the CRG writes and releases position papers regarding new developments, creates model legislation, provides public education, and alerts the public to issues in a variety of areas, including biopiracy, biosafety, and genetically modified foods. The council's concern is that issues regarding genetic technologies remain within the scientific world and do not enter the public realm to be debated or understood. The four principles that the CRG works by are (1) the public must have access to clear and understandable information on technological innovations, (2) the public must be able to participate in public and private decision making concerning technological developments and their implementation, (3) new technologies must meet social needs, and (4) problems rooted in poverty, racism, and other forms of inequality cannot be remedied by technology alone.

One of the CRG's primary strategies is the dissemination of information about genetically engineered foods. As of 2000, there were thirty-five varieties of genetically engineered crops registered with the Food and Drug Administration, the Environmental Protection Agency, or the U.S. Department of Agriculture (Council for Responsible Genetics, 2000). Genetic engineering allows scientists to cross genes from different species together and results in organisms that would never be found in nature.

The CRG's concerns about this technology are based upon a fear that new crops will eventually reduce food security for the world's population because genetic engineering decreases biodiversity and increases the use of pesticides and other environmentally damaging chemicals. The most frequent trait engineered into crops is that of herbicide tolerance. This means that crops can tolerate larger and more frequent applications of chemical herbicides. The ecological and human effects of such chemicals are debated, with some shown to cause birth defects or cancer in laboratory animals.

Another strategy used by the CRG is its Safe Seed Pledge, part of an attempt to educate the public. In 1999, the CRG and a group of seed sellers formulated the pledge, which states that the particular seed company will not knowingly buy or sell genetically engineered seeds. The benefits, the CRG believes, are twofold. First, the companies print the pledge in their catalogs, making the public aware of their conscious practices. Second, the public can learn about the risks of genetically engineered seeds and the benefits of buying organic seeds. The pledge is renewed each year, and in 2001, nearly ninety companies participated (Beland, 2001). Public education is also implemented by the CRG's award-winning magazine, *GeneWatch*. Published since 1983, the bimonthly magazine has focused on such topics as women, reproduction, biotechnology, patents, and genetically engineered food products.

Sources and Further Reading

Beland, Amber. "The Safe Seed Pledge: A Move Towards Food Protection." *GeneWatch*, vol. 14 (2001). http://www.gene-watch. org/magazine. Accessed 21 November 2001.

"Changing the Nature of Nature: An Interview with Martin Teitel." *Multinational Monitor* 21, no. 1 (January–February 2000): 38.

Council for Responsible Genetics. "Frequently Asked Questions about Genetically Engineered Food." (Summer 2000). http://www.genewatch.org/old/website/programs/gefood/faq-food.html. Accessed 21 November 2001.

"DNA Patents Create Monopolies on Living Organizations." (April 2000). http://www.actionbioscience.org. Accessed 23 November 2001.

Teitel, Martin, and Kimberly A Wilson. *Genetically Engineered Food: Changing the Nature of Nature.* Rochester, VT: Park Street Press, 1999.

Jeff DeBonis (1951–)

U.S. Forest Service policies have frequently been criticized by both environmental groups and timber interests but not to the degree they were when Jeff DeBonis started speaking out. In 1989, he formed the Association of Forest Service Employees for Environmental Ethics (AFSEEE). The aim of the association is to encourage change within the Forest Service, especially regarding timber harvests and their environmental consequences.

From 1975 to 1977, DeBonis served as a Peace Corps volunteer in El Salvador; after completing his service, he joined the U.S. Forest Service. In 1978, he was a trainee in the Kootenai National Forest in Montana, where he saw devastating erosion and washouts from timber cutting, with soil draining into nearby trout fisheries. Ten years later, while working in the Willamette National Forest in western Oregon, he saw similar signs of over-harvesting—a bald mountainside that, if logged even more, would disrupt fish habitat and that of the threatened northern spotted owl. He drafted a plan to revise the proposed timber sale for the area, but his plan was rejected. He then leaked the plan to environmental groups and caught the media's attention and agency criticism.

After attending an Ancient Forest Seminar in Eugene, Oregon, in 1989, he reported to his supervisors that the Forest Service was "perceived by the conservation community as being an advocate of the timber industry's agenda. . . . I believe this charge is true. It is time to start perceiving the conservation community as our allies in developing a strategy that will contribute to an ecologically sustainable lifestyle in the 21st century" (McLean, 1990). He gave the letter to agency officials and also distributed it over the Forest

Service's Data General, a computerized communication system. He received about fifty comments on his report and touched off a major controversy.

DeBonis sent an eight-page, single-spaced letter to Chief Forester Dale Robertson explaining his perception of Forest Service practices. Attending a public meeting on old-growth timber in Portland, Oregon, in March 1989, he developed a flyer explaining his views and placed it outside the meeting hall, hoping that other people would respond. He received 200 comments from those who had seen his flyer and from that meeting established the AFSEEE. The organization DeBonis founded calls for "management practices that reflect true stewardship and an ecologically sustainable economic base" (McLean, 1990). The group supports the values of wildlife, fisheries, clean water, and aesthetics on an equal footing with timber sales. It is not an employee union but a group of concerned employees and members of the public seeking forest reform.

He stayed with the Forest Service until February 1990, when—after twelve years of service—he resigned to spend all of his energies on the association and its development. In just a year, he had published two issues of the group's newsletter, *Inner Voice*, sending copies to ranger stations and distributing many to Forest Service retirees and friends. Subsequently, a group of national forest supervisors from across the country sent a memo to Chief Robertson, telling him that some employees, as well as the public, did not view the agency as a leader in conservation. They said they were eager to work with the chief to develop a new role in natural resource conservation issues. These employees tend to be younger than the old guard career foresters. They include scientists and others who believe the forest should be viewed more as an ecosystem and less as a timber company's source of income.

Individuals can join the association anonymously so their jobs will not be threatened. But only about 5 percent of the agency's workers have joined the AFSEEE, and the majority of the members are not Forest Service employees. DeBonis believes many agency staff are subject to intimidation and realizes that some district rangers are critical of his activism against the government. He has been banned from talking to some employees, an action DeBonis takes as a sign the agency believes his organization is effective.

Sources and Further Reading

Breen, Bill. "Free Speech in the Woods." *Garbage* 3, no. 4 (July–August 1991): 16.

Egan, Timothy. "Forest Service Abusing Role, Dissidents Say." *New York Times* (4 March 1990): 1.

Lemonick, M. D. "Whose Woods Are These?" *Time* 138, no. 23 (9 December 1991): 70.

McLean, Herbert E. "A Very Hot Potato." *American Forests* 96, nos. 3–4 (March–April 1990): 30–34.

Phillips, Kathryn. "A Grass-Roots Organization Changes the Fate of the Forest." *Omni* 15, no. 2 (November 1992): 18.

Schubert, Charlotte. "Getting the Ax: Whistleblowers Felled by the Forest Service." *The Progressive* (April 1992): 24–27, 30.

The Downwinders

Hanford, Washington, is home to what is euphemistically referred to as a nuclear reservation. The 500-square-mile area in the eastern part of the state was the source of the plutonium that was used in the atomic bomb that destroyed the Japanese city of Nagasaki at the end of World War II. During the Cold War, additional reactors and processing plants were added, using water from the Columbia River to cool the reactors. The water was then returned to the Columbia River, contaminated by radiation and toxic chemicals. An estimated 440 billion gallons of liquids were also discharged into the ground, contaminating an estimated 200 square miles of groundwater ("A Brief History," 2001).

Hanford is a company town, where virtually everyone is associated in one capacity or another with Rockwell Industries, the defense contractor that managed the facility. During the 1980s, residents began to discover that their community had suffered radiation exposure for decades without their knowledge. As the veil of secrecy that had been in place during the Cold War began to unravel, the secrets of Hanford were exposed, publicized, and documented and eventually became a major environmental issue.

Residents had long suspected that something might be wrong. The incidence of birth defects, thyroid cancer, and deformed animals in the area was gradually being compiled, as neighbors told their stories to one another, and a pattern began to emerge. The

Centers for Disease Control acknowledged that the area's residents were suffering a much higher risk of thyroid cancer than other nearby regions, and reporters, using the Freedom of Information Act, began to get access to previously classified documents.

In order to educate residents, several Hanford-area activists formed the Hanford Education Action League (HEAL) so that those with information about the facility would no longer be afraid to come forward to tell their stories. Among the documents discovered was confirmation of the Green Run—an experiment conducted in 1949 in which the government deliberately released radiation from Hanford to track exposure and releases. The Green Run had been so secret that many people had dismissed it as a myth, unwilling to even contemplate that the government would put innocent civilians at risk of radiation poisoning.

As the number of affected persons began to grow, and their medical problems were confirmed, more than twenty people who had lived in the area in the 1940s and 1950s got together, fueled by anger and distrust. The group became known as the Downwinders, banding together to produce a newsletter and hold meetings where they shared information, despite the lack of a formal organizational structure or agreed-upon goals. They faced hostility from other local residents, who felt that by stirring up old memories, the group would hurt the local economy. Others would still not believe that their government would be a party to such a deliberate act, and some believed that the danger had been grossly exaggerated.

A split caused by internal conflict broke the group into two halves, one operating out of Seattle and the other in eastern Washington. HEAL joined the Downwinders, who also got the support of Native American groups who had also suffered high cancer rates and rheumatoid arthritis. HEAL became the public relations and strategy element, finding foundation support and forming a national lobbying effort. The Downwinders, a research and educational foundation established in 1978 in Salt Lake City, Utah, served as the face of radiation exposure, telling anyone who would listen about their illnesses and seeking compensation from the government.

It was not until 1988 that the federal Department of Energy admitted its role in past experiments, acknowledging that the site was highly contaminated. Six years later, the agency's director admitted that the government might hold some responsibility for the losses of those who had been exposed. But it was difficult for the Downwinders to convince others that there was a cause-and-effect relationship linking Hanford and their medical problems. Hanford became known as a "throw away" community that no one wanted to save.

Congress passed the Radiation Exposure Compensation Act (RECA), with a goal of providing some financial compensation for workers who were exposed. Paired with the Energy Employees Occupational Illness Compensation Program, the federal government began processing claims as the Downwinders continued their fight for "justice for their injuries" ("About Downwinders," 2001). The group has expanded its advocacy agenda to include chemical and biological weapons research, nuclear waste, uranium mining, groundwater issues, and military toxics. Some members and others whose bodies were contaminated have filed lawsuits, but there are an estimated 7 billion pages of government documents that still have not been released. On 27 January 2001, the Downwinders and their supporters commemorated the fiftieth anniversary of atomic testing in Nevada.

Sources and Further Reading

"A Brief History of the Hanford Nuclear Reservation." http://www. downwinder.com/history. Accessed 1 October 2001.

"About Downwinders." http://www.downwinders.org. Accessed 1 October 2001.

Gerber, Michele S. *On the Home Front: The Cold War Legacy of the Hanford Nuclear Site.* Lincoln: University of Nebraska Press, 1992.

Glazer, Penina M., and Myron P. Glazer. *The Environmental Crusaders: Confronting Disaster and Mobilizing Community.* University Park: Pennsylvania State University Press, 1998.

Loeb, Paul. *Nuclear Culture.* Philadelphia: New Society Publishers, 1986.

Rhodes, Richard. *The Making of the Atomic Bomb.* New York: Simon and Schuster, 1986.

Earthjustice

With nine regional offices throughout the United States, an international agenda, and more than thirty years of legal representation, Earthjustice, formerly known as the Sierra Club Legal Defense Fund, has become one of the key nonprofit law firms in the United States. Although the group's primary function is to file lawsuits without charge for public interest clients, it also aims to protect existing environmental statutes from being watered down or overturned. Through the two environmental law clinics it has established at the University of Denver and Stanford University, Earthjustice is also training law students and preparing them to work for other organizations. On 1 June 2000, another public interest firm, Earthlaw, merged its legal staff with Earthjustice's attorneys in a move designed to provide more effective legal representation.

One of the most historically noteworthy actions of the group focused on a development in California's Sierra Nevada range planned by Walt Disney Productions. In one of the first citizen-enforced environmental law cases, *Sierra Club v. Morton*, the U.S. Supreme Court established in 1972 the right of citizens to sue to protect natural resources. The case served as a precedent for other citizen-based suits that have become common strategies for environmental groups.

Typical of other environmental defense groups, Earthjustice has often sued a government agency to enforce a law. In 1968, for instance, the National Park Service was alleged to have failed to protect critical timberland adjacent to the new Redwood National Park. The lawsuit resulted in a doubling of the size of the park as nearby timberland was acquired from logging companies. On the border of Yellowstone National Park, the group successfully sued a mining company that had planned to reopen a gold and silver mine. The company was not allowed to resume mining operations and was fined millions of dollars in penalties, and the site is undergoing restoration.

In recent years, Earthjustice has become more involved in issues related to environmental justice. Citizens in a predominantly minority community in Louisiana sought help from the organization when a uranium plant was scheduled to be constructed in their area. After hearing citizens' protests, the Nuclear Regulatory Commission denied the necessary permit for the facility. The organization has also expanded its agenda to a global perspective, seeking to address human rights and the environment, international trade, and support for environmental legislation in other countries.

Although legal expertise is a valuable resource, the organization also uses its own policy experts who work to prevent legislative backlash and the amending of long-standing environmental statutes. Staff members have focused on the Endangered Species Act and the Clean Air Act, two laws that have been targeted by conservative members of Congress and their supporters. After several years of attempting to develop a strategic plan, the group turned to an advertising agency. The company, Underground Advertising, surveyed potential donors, competing organizations, and prior marketing efforts and developed a new logo, brochures, and a tag line that has become synonymous with the organization's goal, "Because the earth needs a lawyer."

Sources and Further Reading

"About Us: Major Accomplishments." http://www.earthjustice.org/about/major. Accessed 15 October 2001.

Drielak, Steven C. *Environmental Crime*. Springfield, IL: C. C. Thomas, 1998.

Earthwatch Institute

In its mission statement, Earthwatch Institute notes that it "operates on a very simple but radical notion: that if you fully involve the general public in the process of science, you not only give them understanding, you give the world a future." Even though this might at first sound like traditional environmental rhetoric, there is a difference. Earthwatch activists learn by doing.

Founded in 1971 in Boston, the nonprofit group now has over 50,000 members and supporters. Offices in Oxford, England; Melbourne, Australia; and Tokyo, Japan, coordinate Earthwatch activities, which focus on research, conservation, and education. All three goals are met by the organization's unique volunteer program, in which more than 4,000 participate each year. The volunteers, who pay about $1,600 each for the experience, work side by side with the institute's researchers on six continents throughout the world. There are seven main categories of research: endangered ecosystems, oceans, biodiversity, cultural diversity, global change, world health, and archaeology. Scientists may apply for funding through the Center for Field Research, and scholarships are available for high school students and teachers.

One of the group's concerns is that researchers seldom have sufficient resources to conduct expeditions in the field. Many potential projects are abandoned, and even when research is completed, it seldom is reported by the media. By involving citizens in the research projects themselves, Earthwatch believes it is possible to engage the public, bring findings to the policy agenda, and serve a purpose greater than just academic curiosity.

Earthwatch produces an illustrated guide to research projects where volunteers are needed. Some choose to work on a one-week project, whereas others may serve even longer in the field, working with noted scientists and getting hands-on training. Some of the group's members are "armchair explorers" who are able to support the organization's efforts without becoming involved in an expedition. Other volunteers use the field research project as an opportunity to travel, taking vacation time or a summer break to participate. The volunteers themselves come from all over and represent a variety of interests and ages. They may be retirees, artists, nutritionists, health care professionals, teachers, architects, scientists, or simply activists who seek out a different way of expressing their desire to help protect the environment.

More than 2,000 scientific papers and books have been published through Earthwatch research, and hundreds of species new to science have been discovered. Many projects have been conducted in partnership with other groups, such as Conservation International, or have been sponsored by one of the group's corporate sponsors. For example, Ford Motor Company donated $5 million to Earthwatch Institute to implement conservation research stations at key sites of exceptional conservation value that are also highly threatened. The company is also funding a fellowship program to enable educators and staff from corporations to work at the field stations.

Experiential education opportunities such as those provided by Earthwatch are in one sense priceless. Some volunteers stay in European hotels; others sleep in tent camps. But regardless of the nature of the accommodations, participants agree that the experience provides them with the skills and motivation to undertake local community projects when they return.

Sources and Further Reading

Basinger, Julianne. "To Scientists Who Use Paying Volunteers in Fieldwork, the Benefits Outweigh the Bother." *Chronicle of Higher Education* (19 June 1998): A14.

"Ford Motor Company to Donate $5 Million to Earthwatch Institute." http://www.Cartrackers.com. Accessed 8 November 2001.

Gallagher, Leigh. "Walter Mitty Meets Uncle Sam." *Forbes* 167, no. 14 (11 June 2001): 160.

"Welcome to Earthwatch Institute." http://www.earthwatch.org/abouted. Accessed 5 November 2001.

Get Oil Out!

Santa Barbara, California, was one of the first cities where environmental activism began to establish a local presence, with the founding of a grassroots organization known as Get Oil Out! (GOO!). The group was founded as a direct result of the natural gas blowout and subsequent oil spill that occurred on 29 January 1969 six miles off the coastline. The riggers on Union Oil Company's (Unocal) Platform A had drilled a well about 3,500 feet beneath the ocean floor, and during the routine replacement of a drill bit, a pipe burst, triggering breaks in the ocean floor that began to seep oil and gas.

It took oil workers eleven days to cap the rupture, forcing a chemical "mud" down the shaft. In the meantime, an estimated 200,000 gallons of crude oil created an oil slick covering over 800 square miles (Santa Barbara Wildlife Care Network, 2001). Tidal action and winds blew the oil onto a 35-mile stretch of southern California beaches and as far out as the Channel Islands. In an attempt to mitigate the spill, airplanes tried to break up the oil by dumping huge loads of detergent on the ocean waters, and boats used skimmers to try to contain patches of oil. The damage to the coastline brought out thousands of volunteers who spread straw on the sand to soak up the tar and used steam-cleaning equipment on rocks and outcroppings. Other volunteers, many of them students at the University of California, Santa Barbara, worked with the local zoo to rescue injured marine life and seabirds. Many of the animals were poisoned when they ingested the oil or the detergent that had been spread on the slick.

Within days, outraged local residents formed GOO! to protect the region from the negative impact of oil development. Founded by Bud Bottoms, the new group used a series of strategies that were virtually unknown within the environmental community at the time. One tactic was to burn or tear up oil company credit cards in protest; others advocated not purchasing gas from firms associated with offshore drilling. Some early energy conservationists urged the entire nation to reduce its driving instead of boycotting gas stations. GOO! leaders distributed bumper stickers and buttons that read Get Oil Out! Get Alternatives In in a circle with a slash across it.

GOO! activists realized that in addition to the ecological damage caused by the oil spill, the event also would have an impact on the beauty and aesthetic value of the coastline, and that eventually there would be an impact on tourism and the local economy.

Students living in the university's residence halls, many of which were located on cliffs just above the ocean, awoke one day to find that their showers now had gallon-size bottles of solvent to help them clean off the sticky tar that clung to their feet when they walked along the beach. The area's sizable surfing population put up signs warning others of the dangers in the water below. Local hotels reported a downturn in reservations as visitors canceled their plans when they heard about the devastation.

One of the factors that has made GOO! significant in the history of environmental activism is its longevity. Many local groups that form in response to a perceived problem disband and disappear after the problem is solved or when public interest wanes. But GOO! is still alive and well, although it has changed its focus slightly. Building on a coalition of other environmental organizations that did not exist in 1969, GOO! now has a much broader agenda. Because of concerns about human health risks, air pollution, and toxic emissions, GOO! now hopes to get the oil industry not only out of the Santa Barbara Channel or land-based operations nearby but also out of the United States and eventually the entire planet. This ambitious plan stems from a mission statement that includes monitoring legislation and regulations related to oil industry development, encouraging the development of alternative fuels and energy sources, promoting public awareness, and carrying on educational and legal advocacy.

Some observers believe that the 1969 oil spill was the singular event that captured the public's attention about the environment and that GOO! became the model of what local residents can do. It may have served as the element necessary to get Americans involved in the first Earth Day a few months later. After more than thirty years, many of the original activists remain involved, and dedicated, to Get Oil Out!

Sources and Further Reading

Easton, Robert. *Black Tide: The Santa Barbara Oil Spill and Its Consequences.* New York: Delacorte Press, 1972.

"Get Oil Out!" http://www.getoilout.org. Accessed 15 August 2001.

Nash, A. E. Keir, Dean Mann, and Phil G. Olsen. *Oil Pollution and the Public Interest: A Study of the Santa Barbara Oil Spill.* Berkeley: Institute of Governmental Studies, University of California, 1972.

Santa Barbara Wildlife Care Network. "1969 Oil Spill." http://www.silcom.com/~sbwcn.spill. Accessed 15 August 2001.

Steinhart, Carol E., and John S. Steinhart. *Blowout: A Case Study of the Santa Barbara Oil Spill*. North Scituate, MA: Duxbury Press, 1972.

Walstead, Kay. *Oil Pollution in the Santa Barbara Channel*. Santa Barbara: University of California Library, 1972.

Lois Gibbs (1951–)

The name Lois Gibbs is synonymous with New York's Love Canal, although the shy mother from Niagara Falls never imagined herself to be one of the nation's best-known environmental advocates. In 1980, former president Jimmy Carter referred to her as the most important grassroots leader of the Love Canal residents.

Born In Buffalo, New York, Lois Gibbs was one of six children in an abusive family, an experience that led her to promise herself that she would never stay with a husband who would hurt her children. She married Harry Gibbs, a worker at a chemical plant, and together they had two children. In 1972, the family moved to a subdivision called Love Canal, named after the man who had begun a project to connect the upper and lower Niagara Rivers in 1892. The project was never completed, however, so the trench was used as a city dump, as well as by the U.S. Army and the Hooker Chemical Corporation. The site was covered with dirt in 1953, sold to the board of education for $1, and used for a school and homesites.

At the time, Gibbs did not imagine herself as an activist. Her brother was in Vietnam, and she believed that the United States should not be there. But she did not protest the war because she felt it would have been disrespectful to him. In 1978, she found she could not ignore her instinct that something might be responsible for her son Michael's recurring illnesses. In 1976, Calspan, a private research organization, had found chemicals leaking at Love Canal, and two years later, Mike Brown of the *Niagara Gazette* wrote a series of articles about the chemical contamination. Gibbs began going door-to-door with a clipboard and a three-sentence prepared script that she used to ask her neighbors whether or not they were concerned. Somewhat surprisingly, many of them were, citing unexplained family illnesses. Gibbs and two of her neighbors circulated a petition; they took the petition, with 161 signatures, to the New York State Department of Health, and in June 1978, the department held a series of public hearings. Reports of illness and chemical burns were enough to force the department to close the elementary school that had been built on the site in 1955 and to recommend in August 1978 that pregnant women and children be temporarily relocated. The federal government declared Love Canal an emergency area, and Governor Hugh Carey signed an order for the permanent relocation of 239 families known to have had miscarriages and birth defects.

Lois Gibbs organized the Love Canal Homeowners Association at that same time, seeking to have the residents' homes purchased at a fair market value and to deal with others whose homes were nearby but who had not been relocated. Working without a salary, she began testifying at hearings, working with scientists on a comprehensive survey of the residents' health problems, and advocating for relocation funds. New York delayed implementation of the governor's order until 1979, and in 1980, the federal government allocated funds for the relocation. Angry residents, tired of the delays, held a demonstration on 19 May 1980 in which two Environmental Protection Agency workers were held hostage until the government agreed to authorize the permanent relocation of 810 families. Over $15 million was authorized to purchase the abandoned homes; 239 of the ones closest to the canal were demolished, but in 1988, New York officials decided to allow 200 other homes to be reinhabited. Ten years later, of the 800 families, 67 had decided to stay at Love Canal and the other 733 had moved elsewhere ("Love Canal," 2001).

The publicity generated by the incident led Gibbs to expand her activities and to move to Washington, D.C., where she founded the Citizen's Clearinghouse for Hazardous Waste, which became the Center for Health, Environment, and Justice in 1997. Her efforts have led to increased public awareness, and no commercial hazardous waste landfills have been built in the United States since Love Canal. Now based in Falls Church, Virginia, the organization has a $1 million annual budget, 27,000 members, and a coalition of contacts with 10,000 grassroots groups ("Lois Gibbs, Champion of Love Canal," 2001).

Sources and Further Reading

Brown, Michael. *Laying Waste*. New York: Pantheon Books, 1979.

Gibbs, Lois. *Love Canal: My Story*. Albany: State University of New York Press, 1982.

———. *Dying from Dioxin*. Boston: South End Press, 1995.

Hofrichter, Richard, ed. *Toxic Struggles: The Theory and Practice of Environmental Justice*. Philadelphia: New Society Publishers, 1993.

Levine, Adeline Gordon. *Love Canal: Science, Politics, and People.* Boston: Lexington Books, 1982.

"Lois Gibbs, Champion of Love Canal, Working to Rebuild Democracy." http://www.netaxs.com. Accessed 28 September 2001.

"Love Canal." http://history.sandiego.edu/gen/nature/lovecanal.html. Accessed 28 September 2001.

Wayne Hage (1936–)

It seems sometimes that Wayne Hage has spent most of his life in court fighting the federal government. His form of environmental activism stems from his background as a biologist, rancher, author, and leader in the property rights movement. He has waged his own kind of war on the Bureau of Land Management, the Department of the Interior, the U.S. Forest Service, and other federal and state agencies that he believes have been captured by environmental groups and their leaders. In the editor's introduction to the third edition of Hage's book, *Storm over Rangelands,* Ron Arnold writes that the book "has swept away a century of ignorance and misinformation about our federal lands . . . and America will never be the same again" (Hage, 1994, x). Hage's critics would most likely agree, but for different reasons.

In 1978, Hage's family of seven purchased the Pine Creek Ranch in a remote section of central Nevada, adjacent to the Toiyabe and Humboldt National Forests. The ranch combined four original homesteads, and the family assumed that ranch ownership included vested water and grazing rights on their 7,000 deeded private acres. In 1979, Forest Service officials told Hage that they were surveying the area for water so that the government could file a claim on all the water in the Monitor Valley. After contacting the Nevada state engineer's office, Hage learned that the federal government had included in its claim water on 160 acres of his ranch.

In 1981, Hage filed a petition with the state engineer for a determination of who actually owned the rights in the valley, a process that would taken ten years instead of months. Meantime, his ranching operation was on the brink of financial collapse. He then filed a suit in the U.S. Court of Claims in September 1991, arguing that under the Fifth Amendment to the Constitution, he was entitled to compensation because the government had "taken" his land without compensating him. He believed that the government owed him for the land and water rights, an irrigation ditch right-of-way, forage rights, rangeland improvements, and cattle. In response, Hage and an employee were charged with cutting and removing juniper brush from an irrigation ditch—a felony the government charged as taking government property. The felony charges were later dismissed by the Ninth Circuit Court of Appeals.

The U.S. Forest Service produced photographs it claimed showed overgrazing on Hage's ranch, and the agency canceled his grazing and water permits for five years. Hage said the bare land was due to the natural cycle of grass growth, which returned that spring. He also received a call from a Toiyabe National Forest official saying that some of his 2,000 head of cattle were trespassing on government land. Hage was in the process of moving the herd from winter to summer grazing areas and said it was not unusual for some of the cattle to stray. Armed federal agents confronted him on several occasions, and in July 1991, more than 100 cattle were confiscated and later sold at auction. The Forest Service kept the profits from the sale and then sent Hage a bill for the confiscation costs.

Environmental groups, including the Sierra Club, the National Wildlife Federation, and the Natural Resources Defense Council, filed to serve as intervenors in the suit, *Hage v. United States*, arguing that Hage should not receive compensation for losing the permits. The Court of Claims denied the groups' motion, and in March 1996 handed down a ruling that allowed Hage's takings claim to proceed to trial. At the same time, CIGNA Corporation, which held the mortgage on the Hage ranch, announced it was going to sell the property at auction. Helen Chenoweth, a member of the Idaho congressional delegation, heard about Hage's problems and called a company representative into her Washington office. She accused CIGNA of possibly violating conspiracy laws by contributing to the National Wildlife Federation, which was alleged to be working with the Forest Service to take Hage's property. Chenoweth threatened to call for an immediate Justice Department investigation and followed through on the House floor. Shortly thereafter, Hage received a call from CIGNA telling him the sale had been canceled. He and Chenoweth later married.

In November 1998, the court gave a preliminary opinion in *Hage v. United States,* and the parties are now awaiting the final decision on the property rights phase of the trial. When the court rules on the rights owned by the plaintiffs, it could advance to a decision on whether or not the government's action involved a taking. Hage's daughter, Margaret Hage Gabbard, says the case is about Americans regaining control of their government, adding, "If Americans are to survive the rabid environmental agenda, we must protect property rights" (Gabbard, 2001).

Sources and Further Reading

Cawley, R. McGreggor. *Federal Land, Western Anger: The Sagebrush Rebellion and Environmental Politics.* Lawrence: University Press of Kansas, 1993.

Echeverria, John D., and Raymond Booth Eby. *Let the People Judge: Wise Use and the Private Property Rights Movement.* Washington, DC: Island Press, 1995.

Gabbard, Margaret Hage. "Restoring Americans' Property Rights: *Hage v. U.S.*" http://www.stewardsoftherange.org/news.htm. Accessed 30 December 2001.

Hage, Wayne. *Storm over Rangelands.* 3d ed. Bellevue, WA: Free Enterprise Press, 1994.

Yandle, Bruce, ed. *Land Rights: The 1990s' Property Rights Rebellion.* Lanham, MD: Rowman and Littlefield, 1995.

Denis Hayes (1944–)

Denis Hayes has been called one of the 100 most influential Americans of the twentieth century. He has hitchhiked across Southeast Asia, Africa, Europe, and the Middle East. He was awarded the Jefferson Medal by the American Institute for Public Service for the greatest public service by an American under age thirty-five. He has been director of a national laboratory and an environmental foundation. He enjoys tropical reef diving and whitewater rafting. He has been an adjunct professor at Stanford University. Denis Hayes says he is not sure if he has a profession, but he is one of the youngest and best-known figures to emerge from the environmental activism of the 1970s.

Growing up in the Pacific Northwest, Denis Hayes developed a love for the natural world near his home in Camas, Washington. After attending Clark College for two years, he decided that his real passion lay in seeing the world. When he returned to college as a history major at Stanford University, his travel experiences clearly colored his activism, initially in opposition to the war in

Vietnam. He left the West to attend Harvard, where his life took a dramatic turn in 1969.

Senator Gaylord Nelson of Wisconsin had been trying for months to build environmental awareness in the United States at a time when antiwar protests seemed to be a natural part of university life. He saw numerous teach-ins where faculty and students would spend hours discussing the war and thought that perhaps the same idea could be used to spur environmental activism. Nelson turned to Hayes to coordinate his still-incomplete idea.

With a budget of $190,000 (including some of Nelson's personal funds), Denis Hayes took on the task of organizing the first Earth Day on 22 April 1970. He took out an advertisement in the *New York Times* announcing the plans for the events and hoping to encourage college students to participate. By building a network of supporters and activists from college campuses across the country, he developed a base of support that would later bring together millions of people around the world. The day's events were indeed historical. John Lindsay, mayor of New York, closed Fifth Avenue to all traffic. There was an entire week of speakers and events in Philadelphia and symbolic protests in San Francisco.

In 1973, Hayes became the director of the Solar Energy Research Institute during the Carter administration as an advocate for alternative sources of energy. When President Ronald Reagan cut funding for the project in 1975, Hayes went back to college, returning to Stanford, receiving a law degree, and becoming an adjunct professor of engineering there from 1983 to 1989.

Twenty years after the original observance of Earth Day, Hayes came back to the environmental movement to organize the 1990 celebration. That year, more than 141 countries were involved, with an estimated 200 million participants around the world ("Denis Hayes," 2001).

From there, he moved to Seattle and has built upon his environmental career by serving as the president of the Bullitt Foundation, as chairman of the board of Green Seal, as head of the Earth Day Network, and as chairman of the board of the Energy Foundation.

Sources and Further Reading

"Denis Hayes." http://www.earthday.net. Accessed 28 September 2001.

"Denis Hayes." http://www.library.thinkquest.org. Accessed 28 September 2001.

Hayes, Denis. *Rays of Hope: The Transition to a Post-Petroleum World.* New York: W. W. Norton, 1977.

———. *Pollution: The Neglected Dimensions.* Washington, DC: World watch Institute, 1979.

Lowery, Linda. *Earth Day.* Minneapolis, MN: Carolrhoda Books, 1991.

Oelschlaeger, Max. *After Earth Day.* Denton: University of North Texas Press, 1992.

Velma B. Johnston (1912–1977)

As a child growing up on her parents' ranch in rural Nevada, Velma Johnston learned a lot about horses from her father. She was taught that the best way of training them was through gentle methods, rather than "breaking" them. One of four children, she contracted polio when she was eleven years old and spent much of her time studying and working with animals on the ranch.

In 1950, Velma made a discovery that would change her life forever. Driving behind a truck hauling horses, she noticed blood dripping out and followed the vehicle to a rendering plant. While hiding, she saw that the horses in the truck were packed so tightly together that they could not move. What caught her attention was the sight of a yearling horse that was being crushed by the weight of two stallions. Although she had heard that mustangs were being herded by airplanes and then sold for slaughter, she had had no idea of the cruelty of the wild horse roundups. The animals were chased by the planes until their strength gave out, hastily herded together, and placed on trucks or trains. Many died because of a lack of food or water or because of injuries they had sustained.

Two years later, the Reno secretary joined in a petition drive in Storey County, Nevada, to stop the use of airplanes to herd wild horses for eventual slaughter. Johnston presented vivid photographs of the cruelty, and the county's board of commissioners agreed to a ban on the use of the planes. She became an activist and advocate, drawing support from animal lovers all over the United States. In 1955, a critic in the Nevada legislature referred to her derisively as "Wild Horse Annie," but she gained passage of state legislation similar to the Storey County regulations. She began a campaign in Congress to prevent the inhumane capture and treatment of wild horses and burros, supported in large part by thousands of children who wrote letters asking how people could be so cruel to the animals.

Congress responded in 1959 by enacting Public Law 86-234, known as the Wild Horse Annie Law. But there was little enforcement of the ban on the use of motorized vehicles to round up wild horses, and in 1971, she and other advocates convinced Congress to pass an even more substantial bill, Public Law 92-195, the Wild Free-Roaming Horse and Burro Act. She founded the group Wild Horse Organized Assistance (WHOA!) to save the diminishing herds of wild horses in the West. Her efforts garnered a public service award from Interior Secretary Rogers Morton in 1972, but she was also threatened by an Idaho vigilante group who opposed her campaign. Velma Johnston died of cancer in 1977 and is remembered in a monument at Indian Park in the Little Bookcliffs Herd Management Area in Colorado, the site of a herd of over 100 wild horses.

Sources and Further Reading

Henry, Marguerite. *Mustang, Wild Spirit of the West.* Chicago: Rand McNally, 1966.

"The Mustang Dilemma." Public Broadcasting System. http://www.pbs.org/wildhorses. Accessed 21 September 2001.

"Velma B. Johnston." Equinenet. http://www.equinenet.org/life/whannie.html. Accessed 23 September 2001.

Wild Horse and Burro Freedom Alliance. http://www.savewildhorses.org/annie.htm. Accessed 21 September 2001.

"Wild Horses of the Pryor Mountains." http://www.webcom. com/~ladyhawk/welcome. Accessed 21 September 2001.

Robert F. Kennedy Jr. (1954–)

The Kennedy name has always opened closed doors and given its namesakes credibility. Although some Kennedys are more associated with scandal than the environment, one particular member of the family has made his name synonymous with environmental protection. As nephew of President John F. Kennedy, and the son of the former presidential candidate and attorney general who was assassinated in 1968, there may have always been high expectations for the attorney cum environmental activist. But there is little doubt among those who know him that this Kennedy sincerely believes in his work and his goals.

After his graduation from Harvard University, Robert F. Kennedy Jr. studied at the London School of Economics and received a law degree from the University of Virginia. After his graduation from law school, he did graduate work at the Pace University School of Law and received a master's degree in environmental law. He worked as an assistant district attorney in New York and helped his uncle, Edward M. Kennedy, in his 1980 presidential campaign.

Later, Robert F. Kennedy Jr. became the senior attorney for the Natural Resources Defense Council, one of the nation's most respected public interest law firms. He also serves as the chief prosecuting attorney for Riverkeeper, Inc., an organization founded in 1983 by members of the Hudson River Fisherman's Association to monitor and protect the region's watershed.

In recent years, Kennedy has expanded his interests to cover a broad range of political and environmental issues. He is credited for having engineered a 1997 watershed agreement to regulate development around the reservoirs that provide drinking water for New York City. In 1999, he accused popular New York Mayor Rudolph W. Giuliani of putting his political ambitions ahead of the protection of the city's drinking water by failing to enforce the agreement. In a report issued by the environmental group Riverkeeper, Inc., Kennedy also said the city's Department of Environmental Protection was becoming "an agent of destruction in the New York City watershed." Giuliani responded that the report covered previously identified problems, and the deputy mayor called it "a cheap political document focusing on his public relations and not substance" (Bumiller, 1999). Kennedy supported Giuliani's opponent in the 2000 race for the U.S. Senate, former first lady Hillary Rodham Clinton, who eventually won the New York election. At one time, Kennedy considered a run for the Senate himself.

In 2001, Kennedy was sentenced to thirty days in jail for his participation in a protest over Navy bombings at Camp Garcia on the island of Vieques. He had joined a number of celebrity activists who called for an end to the bombing maneuvers, which were believed to have caused negative environmental effects. Although he had been represented by former New York governor Mario Cuomo, whose son Andrew is Kennedy's brother-in-law, his defense of having been involved in the tradition of civil disobedience did not sufficiently capture the support of the judge. But the high-profile demonstrations were sufficient to persuade President George W. Bush to order a halt to the island bombings by 2003.

During the 2000 presidential election, Kennedy was highly critical of Ralph Nader, who ran as the Green Party candidate for president. In an opinion piece published in the *New York Times*, Kennedy accused Nader of potentially torpedoing efforts to address the nation's most important environmental challenges. He wrote that Nader would siphon votes away from the Democratic challenger, Vice President Al Gore—a prediction that was fulfilled later on when Gore narrowly lost the election.

Kennedy has subsequently addressed issues ranging from the factory hog industry and Clean Water Act violations to a luxury housing development in Clifton Cay in the Bahamas. He has spoken to large groups such as the American and Canadian Lung Associations and the International Association of Therapeutic Drug Monitoring and Clinical Toxicology.

He has used his celebrity status to advance the causes in which he is most active. In 1999, he borrowed seed money from friends and relatives to start Tear of the Clouds, a company that produces a Tiffany-designer bottle for Keeper Springs water, named after the "keeper" groups across the country. A New York advertising firm developed the bottled water's marketing campaign, which does not feature Kennedy himself. The competitive market is crowded with more than 500 brands worth an estimated $4.3 billion in sales; the new company's profits will help finance efforts to preserve and protect the nation's waterways.

His efforts have not been without criticism, however. In June 2000, eight members of the board of Riverkeeper resigned after they learned about an employee that Kennedy had hired to assist in the monitoring of the New York watershed agreement. Riverkeeper founder Robert Boyle said he was appalled that Kennedy had hired William Wegner, who had pled guilty to smuggling rare bird eggs into the United States from Australia and had been convicted of tax evasion in 1999. Boyle fired Wegner, but Kennedy insisted that he be retained, leading to the resignations. Amid all the criticism, he is a licensed master falconer and has written three books.

Sources and Further Reading

Bumiller, Elisabeth. "Robert Kennedy Says Mayor Plays Politics with Water." *New York Times* (10 November 1999): B3.

Cronin, John, and Robert F. Kennedy Jr. *The Riverkeepers*. New York: Simon and Schuster, 1997.

Kennedy, Robert F., Jr. "Nader's Threat to the Environment." *New York Times* (10 August 2000): A21.

Wilkinson, Alec. *The Riverkeeper.* New York: Knopf, 1991.

League of Conservation Voters

In 1969, just before the signing of the National Environmental Policy Act and Earth Day, Marion Eddy was working in Washington, D.C., after recently graduating from college. She discovered that although there was a significant amount of environmental concern developing in the United States, activists had not been able to turn their fervor into electoral success in Congress. A year later, Eddy founded one of the nation's most respected nonprofit environmental organizations, the League of Conservation Voters (LCV).

The LCV has become known as the political arm of the environmental movement, providing an authoritative source of information for voters who want to know where a candidate stands on the issues. In 1970, the group produced its first report, the *National Environmental Scorecard,* which has become the model for other activist groups. The scorecard is developed in consultation with volunteer representatives of twenty-five environmental organizations who analyze the voting records of members of Congress on environmental health and safety, resource conservation, and spending for environmental programs. Members of Congress are rated on a scale from 0 to 100, with the highest number reflecting representatives with the strongest environmental records according to LCV standards. The results are then distributed to the media and to voters throughout the country. The group also works between election cycles and on state ballot initiatives related to environmental protection.

The organization claimed its first victory in 1972 when alerts to grassroots organizations enabled them to unseat two powerful congressional candidates with a record of voting against environmental issues. Eddy stepped down as president in 1985, and the organization chose Alden Meyer as its leader. This was a significant leadership crossroads for the LCV as the group took on the Reagan administration and its supporters in Congress. In 1987, the organization changed strategies and increased the number of political endorsements it issued through its Action Fund. Three years later, Bruce Babbitt, who would later become the secretary of the Interior in the Clinton administration, helped to give the LCV more visibility and credibility. But in 1994, a Republican tidal wave led to the loss of some the most active supporters of the environment in Congress, and in the following two years, a legislative agenda resulted that was designed to roll back much of the progress that had been made in the previous decade.

In 1996, with Deb Callahan serving as president, the LCV undertook its first Dirty Dozen campaign. The LCV identified members of Congress who they perceived to be vulnerable in their bid for reelection and with the worst records involving environmental issues, along with its Perfect Ten EarthList of the strongest environmental candidates. Spending $1.5 million campaigning, seven of the twelve were defeated that year, in part owing to the independent campaigns launched by the LCV. By targeting specific candidates and states, the LCV showed its political clout, hiring field organizers, sending out over a quarter million pieces of mail, running 9,000 radio and television ads, and holding media tours. An additional $450,000 was raised for the Perfect Ten candidates (LCV Annual Report, 1996, 6–7). Two years later, $2.3 million was spent against the Baker's Dirty Dozen (thirteen candidates), nine of whom were defeated. All ten members of the EarthList won in their elections (LCV Biennial Report, 1998, 2). By 2000, LCV leaders moved a step further by publishing presidential profiles that outlined the candidates' environmental records.

The league's board of directors represents a Who's Who of environmental organizations and leaders, with the 2000 list including the group Environmental Defense as well as mainstream groups such as the Sierra Club, Defenders of Wildlife, the Izaak Walton League, and the Natural Resources Defense Council. The press conference to announce the scorecard results is watched closely by other organizations as well as potential candidates, who recognize that the LCV may also be the voice of other critical groups as well. Although other organizations now have their own versions of a scoring system, including groups whose rankings are virtually opposite that of the LCV, it adds to its reputation by not only analyzing voting results but also working toward the defeat of those with the weakest environmental records. By using celebrities such as Robert Redford to record ads for individual pro-environmental candidates, issues related to the environment can be placed on a political agenda when they otherwise might have been ignored.

Sources and Further Reading

Duran, Nicole, and Alan Greenblatt. "An Abundance of Rankings." *CQ Weekly* (12 August 2000): 1961.

League of Conservation Voters. 1996 Annual Report. Washington, DC: League of Conservation Voters, 1996.

———. Biennial Report: 1997–98. Washington, DC: League of Conservation Voters, 1998.

———. *National Environmental Scorecard 2000*. Washington, DC: League of Conservation Voters, 2000.

Toner, Robin. "Interest Groups Take New Route to Congressional Election Arena." *New York Times*, 20 August 1996.

Living Lands and Waters, Inc.

Imagine 1,667 feet of barge rope, 4,924 car tires, 1,636 steel drums, 350 propane tanks, 287 refrigerators, 75 water heaters, 1 barge, 46 washing machines, and 3 tractors. Add 1 prosthetic leg, a hog-feeding trough, bowling pins, discarded portable toilets, 72 televisions, 49 sinks, 11 bathtubs, 5 motorcycles, and 40 barbecue grills.

These are all items that Chad Pregracke and the volunteers of Living Lands and Waters have gathered along the Mississippi River from 1997 to 2001. Based in East Moline, Illinois, the group is dedicated to picking up the garbage that lines the great river's shorelines, focusing on the northern segment from Guttenberg, Iowa, to St. Louis, Missouri. The group began when Pregracke was a teenager, and he and his older brother Brent, who worked as commercial clammers, camped on the many islands in the Mississippi. Starting with a goal of 100 miles of the river, Chad Pregracke gathered funds from corporate sponsors such as Alcoa Mill Products and state and federal agencies. The vice president of Alcoa in Bettendorf, Iowa, helped the group formulate a business plan, set up as a nonprofit organization, and develop educational programs for elementary school children. The company has also provided nearly $100,000 in donations. Other sponsors such as Anheuser-Busch Companies, Enron, and Caterpillar Tractors have also signed on, as has the Sierra Club and the U.S. Army Corps of Engineers.

During the first summer of operation, Pregracke lived on a houseboat named *The Shanty* and picked up over 45,000 pounds of trash as part of the Mississippi River Beautification and

Restoration Project. The following year, the group purchased a boat to expand their efforts, working seven days a week along a 435-mile stretch of the river. One of many community cleanup efforts enlisted the help of over 170 volunteers who cleaned over 40 miles of shoreline in a few hours. The debris is sorted and put into piles, with as much as possible being recycled and the rest hauled to landfills. Some of the recyclable items, such as propane gas containers, can be sold to help raise additional funds. Pregracke also planned a two-day music event and environmental exposition called River Relief '98 in Davenport, Iowa, which unveiled the national Adopt-a-Mississippi River Mile Project. More recently, the group has contributed to cleanup efforts on the Illinois and Ohio Rivers and is looking to a new project on the Hudson River and stream bank restoration. The Mississippi River itself is over 2,300 miles long.

By August 2001, the organization had a budget of just over $300,000, with six employees, three boats, two barges, and about twenty-five corporate sponsors (Cohen, 2001). The crew, whose members are paid $50 a day, changes often because the work is difficult and dirty. Fighting mosquitoes, heat, and mud, they also must fight their way through stinging nettle and snakes. One volunteer lasted a day.

Sources and Further Reading

Charles, Nick, and Kristin Baird Rattini. "Mr. Cleanup." *People* 54, no. 1 (3 July 2000): 113.

Cohen, Sharon. "He Feels Good." http://www.stacks.msnbc.com. Accessed 18 October 2001.

Duignan-Cabrera, Anthony, and Michael McLaughlin. "Mississippi Cleanup." *Life* 21, no. 8 (July 1998): 78.

Kirn, Walter. "Meet the New Huck." *Time* 156, no. 2 (10 July 2000): 70.

Simon, Stephanie. "Wresting Refuse from the Mississippi." *Los Angeles Times* (15 October 2001): A14.

Donella Meadows (1941–2001)

The name Donella Meadows is synonymous with sustainability— a scientist turned grassroots activist whose work on computer models of world population trends was coupled with training activists in Hungary, Costa Rica, Singapore, Kenya, and Portugal. When she died of bacterial meningitis in 2001 at age fifty-nine,

Meadows had received a prestigious MacArthur Fellowship, had been selected as one of ten Pew Scholars in Conservation and the Environment, and had been nominated for a Pulitzer Prize for her syndicated weekly newspaper column, "The Global Citizen." She referred to herself as "an interactivist, a visionary, a learner, a radical" and as "a farmer and a writer" (Meadows, 1999). Her mother, Phoebe Quist, called her daughter an "earth missionary" ("Obituary," 2001).

Dana, as she was known, was born in Elgin, Illinois, and earned her B.A. in chemistry from Carleton College in 1963. Her training continued at Harvard University, where she received a Ph.D. in biophysics in 1968, going on to become a professor of environmental studies at Dartmouth College in 1972. She left Dartmouth in 1983, although she maintained a role as an adjunct professor, to focus on her writing. She is perhaps best known as a coauthor of *The Limits of Growth,* a 1972 report that exposed the trends of consumption in global resources, using computer modeling and system dynamics to predict future outcomes. The report evolved into a book that sold 9 million copies and was translated into more than twenty languages. The authors prepared twelve paths or scenarios of choice, based on the computer models, which were potential outcomes for the human economy up through the year 2100.

Twenty years later, Meadows and her colleagues updated their work to run their model again. In *Beyond the Limits,* they concluded that consumption was growing more rapidly than they had predicted and that new problems, such as the destruction of the ozone layer, had unexpectedly appeared. Both books were soundly criticized for what some called a "doomsday prediction" that did not take into account technological advances that could solve the problems they identified. She herself criticized social system modeling in a book published in 1985.

Meadows's interdisciplinary approach to the environment is exemplified by the many projects in which she participated. She helped found the International Network of Resource Information Centers to bring together researchers in fifty countries to promote sustainable resource management. She worked as an adviser to the Public Broadcasting System's Boston affiliate to produce an acclaimed ten-part series, *Race to Save the Planet,* from 1988 to 1990. She served on the boards of directors of several organizations, including the Hunger Project, the Trust for New Hampshire Lands, and the Center for a New American Dream. At one point in her career, she criticized the jeans ads used by designer Calvin Klein because they featured teenagers in seductive poses, and she sought support for wider access to birth control and campaigns against irresponsible sex.

She lived her life in concert with her views on sustainability. For twenty-seven years, she lived on a small, communal farm in New Hampshire dedicated to organic management, and in 1999, she founded an "eco-village" in Cobb Hill, Vermont. The Sustainability Institute she established in 1997 is not a think tank but what she referred to as a "think-do-tank" (Smith, 2001). Her legacy there will continue, supported by those who will continue her activism and research.

Sources and Further Reading

Bruckmann, Gerhart, Donella Meadows, and John Richardson. *Groping in the Dark: The First Decade of Global Modeling.* New York: Wiley, 1982.

Meadows, Dennis L., Donella Meadows, and Joergen Randers. *Beyond the Limits: Confronting Global Collapse, Envisioning a Sustainable Future.* Post Mills, VT: Chelsea Green, 1992.

Meadows, Donella H. *The Global Citizen.* Covelo, CA: Island Press, 1991.

———. "Chicken Little, Cassandra, and the Real Wolf—So Many Ways to Think about the Future." *Whole Earth* (Spring 1999): 106.

Meadows, Donella H., et al. *The Limits of Growth: A Report for the Club of Rome's Project on the Predicament of Mankind.* 2d ed. New York: Universe Books, 1972.

"Obituary. 2 February 2001—Donella Meadows." Personal communication to author, 2 February 2001.

Smith, Gar. "Donella Meadows." *Earth Island Journal* 16, no. 2 (Summer 2001): 48.

Chico Mendes (1944–1988)

There are few martyrs in the world of environmental activism, but the murder of Chico Mendes on 22 December 1988 led to an international outcry and a worldwide refocusing on the protection of the Amazon. As one of the first guardians of the rain forest, the Brazilian leader of the *seringueiros,* or rubber tappers, helped bring an end to government grazing subsidies that had led to the destruction of thousands of acres of land in the fragile ecosystem.

Chico Mendes's father, Francisco Mendes, came to the remote Amazon forest near Bolivia and Peru in 1926 after leaving the poverty of his home in Ceara, Brazil. His wife, Iraci Lopes Filho, was a member of a family who for generations made their meager living collecting the white, milky latex from the trees in the Amazon forest. Their son, Francisco "Chico" Alves Mendes Filho, was born in the western Amazon village of Xapuri and at age nine became a rubber tapper, illiterate and poor like his father. At the time, landowners did not allow their workers to build or attend schools, so his education was an informal one. A political refugee, Euclides Fernandez Tavora, worked with him to learn how to read and write, using old magazines and a short-wave radio.

Chico and his family lived in the seringal Cachoeira region, and in the 1970s, he became a leader in a nonviolent resistance movement to defend their homes from cattle ranchers, who demanded that they leave. Along the western border of Brazil, shared with Peru and Bolivia, the government began its National Integration Program, hoping to colonize the region with cattle ranchers and forcing the native people to relocate. Over the next fifteen years, the ancient forests were intentionally burned to make way for farms and ranches, resulting in massive erosion and a loss of jobs.

Mendes gathered his fellow workers together to protest the relocation, organizing blockades against bulldozers. He founded a trade union in Acre, Brazil, in 1975 and the Workers' Party four years later. In 1985, he expanded the movement by creating the National Council of Rubber Tappers and, along with Brazilian anthropologist Mary Helena Allegretti, organized the first national meeting of rubber tappers in Brasilia. Mendes sought aid from environmental groups in the United States to help build rubber tree preserves that would provide the local people with a source of income by practicing sustainable agriculture. Gaining international attention, he was awarded the Global 500 prize by the United Nations Environment Program in 1987 and the Ted Turner Better World Society Environment Award, although he was ignored by the Brazilian government and media. His major success, however, was a winning effort to stop the ranchers from cutting down a forest that the rubber tappers wanted to make into a protected reserve. He built a coalition between the rubber tappers and the indigenous Indians, and his leadership and power became a threat to local ranchers.

He married his childhood friend, Ilzemar Gadlha Bezerra, in 1983, and they had a son and daughter. Throughout 1988, the Mendes family received several *anuncios*, a Brazilian death warning, including one on his forth-fourth birthday on 15 December. A week later, Mendes was assassinated just outside his home while his wife and children watched nearby. Guards who had been hired by the family for protection suddenly disappeared, and it was not until the international environmental community mobilized that the Brazilian government began to investigate the murder.

One of the local ranchers who had threatened Mendes, Darci Pereira, was eventually arrested and confessed that he had shot Mendes, and his father, Darli Alves da Silva, was convicted of plotting the murder. In 1990, Pereira and his father were sentenced to nineteen years in prison. Two years later, da Silva was retried after a judge ruled the prosecution's main witness was biased. In February 1993, Pereira, his father, and seven other inmates escaped from a prison in Rio Branco, Brazil, by sawing through the bars of a window in their cell. Authorities admitted that security at the prison was extremely lax and led a search for the escaped men. The attorney for Mendes's wife blamed authorities for the escape, adding that human rights activists were aware of the men's plans. Darli Alves da Silva was recaptured by Brazilian federal police in June 1996, but his son, Darci, is still at large. Mendes's wife, Ilzemar, became president of the Chico Mendes Foundation, created to help the remaining rubber tappers who still live in the forest. But his death left a void that has remained unfilled because there is no longer a charismatic leader for the union and the workers.

Sources and Further Reading

"Chico Mendes—Resources." http://www.global500.org. Accessed 30 December 2001.

"The Chico Mendes Sustainable Rainforest Campaign." http://www.environmentaldefense.org/programs. Accessed 23 December 2001.

Cowell, Adrian. *Decade of Destruction: The Crusade to Save the Amazon Rain Forest.* New York: Henry Holt, 1990.

Hyman, Randall. "Rise of the Rubber Tappers." *International Wildlife* 18, no. 5 (1988): 24–28.

Revkin, Andrew. *The Burning Season: The Murder of Chico Mendes and the Fight for the Amazon Rain Forest.* Boston: Houghton Mifflin, 1990.

People for the Ethical Treatment of Animals

Supermodels Christy Turlington and Tyra Banks are supporters of People for the Ethical Treatment of Animals (PETA). So, too, are actors Alicia Silverstone and Kim Basinger and directors Martin Scorsese, Oliver Stone, and Rob Reiner, along with musicians Paul McCartney, Belinda Carlisle, and k.d. lang. Tennis star Martina Navratilova and actor Jennie Garth have appeared in support of its campaigns, and Belinda Carlisle, the B-52s, and Indigo Girls have participated in concerts and recording sessions on is behalf.

PETA has developed an unusual base of support for its efforts, which are conducted on a number of political fronts. The organization, formed in 1980, includes among its activities efforts to stop the use of animals in laboratory experiments, the Cut Out Dissection project in schools, a Live and Let Live vegetarian campaign, Rock Against Fur benefit concerts, two record albums *(Tame Yourself* and *Animal Liberation)*, and a mailing on animal rights to an estimated 9 million elementary school children.

Animal rights issues have prompted the development of numerous organizations around the world, but none has the reputation for media coverage, investigative work, protests, and consumer action that PETA has. Even its critics agree that PETA activists have a unique way of garnering attention for their cause. Secretary of Agriculture Dan Glickman was the target in 2000 when a PETA member threw a tofu pie in his face as he spoke at a nutrition meeting in Washington, D.C. Glickman successfully ducked the pie; the protester was photographed being handcuffed and led away, getting the group more publicity. PETA ran an ad in school newspapers in the Midwest that showed a teenager with a milk mustache in a parody of the dairy industry's Got Milk? ads. The PETA ad's headline was Got Zits? and linked animal fats and hormones in milk to acne. A similar campaign was waged in college newspapers using the headline, Got Beer?, but the ads were dropped after the group Mothers Against Drunk Driving (MADD) protested that PETA was encouraging alcohol use with the advertisement.

Sources and Further Reading

Francione, Gary L. *Rain without Thunder: The Ideology of the Animal Rights Movement.* Philadelphia: Temple University Press, 1996.

Guillermo, Kathy Snow. *Monkey Business: The Disturbing Case That Launched the Animal Rights Movement.* Washington, DC: National Press Books, 1993.

People for the Ethical Treatment of Animals. http://www.peta.org.

Strand, Rod. *The Hijacking of the Humane Movement.* Wilsonville, OR: Doral Publishing, 1993.

Wenzel, George. *Animal Rights, Human Rights.* Toronto: University of Toronto Press, 1991.

Ted Turner (1938–)

He had a crazy idea. He would challenge the major television networks by sponsoring a television network that would feature nothing but news, twenty-four hours a day. The network would be utilized to foster social change, especially issues related to the environment. He would fund the production of documentaries, by himself, if necessary.

Ted Turner has done all these things. A self-made millionaire and entrepreneur, he believes in putting his money where his mouth is, and much of his attention is focused on protecting the environment. Millions of dollars are donated to nonprofit environmental organizations through the Turner Family Foundation, which he created in 1991. He is the country's largest private landowner, with holdings of over a million acres in six states and in Argentina. He has pledged that none of the land will ever be developed, and he has promised to help restore the acreage to its original state to try to recover the natural ecosystem.

Few people would have predicted that Ted Turner would do any of these things. Growing up in Cincinnati, Ohio, he attended a boarding school and was "instructed by wire coat hangers" by his father, a man who sold billboard space for an advertising company, then volunteered for the U.S. Navy. He later went to two military academies, briefly attended Brown University (his father berated him for majoring in classics; he changed his major to economics), and served with the U.S. Coast Guard. In 1960, he returned to work in his father's business, the Turner Advertising Company. When his father committed suicide in 1963, Turner took over the company and began to expand its operations, purchasing television stations and turning a profit.

Turner was one of the first entrepreneurs to see the future of cable television using space satellites. In 1976, he used his Atlanta station as a base for what would later become the Turner Broadcasting System (TBS) and then launched Cable News Network (CNN) in 1980. More enterprises and innovations followed, from the founding of Turner Classic Movies in 1994 to the popular television series, *Earthwatch.*

Although most environmental activists rely upon motivation and dedication (and little else), Ted Turner does not only that but is also able to use his personal fortune. He has developed a global platform for his ideas, creating groups such as the Better World Society in 1985 and a children's cartoon program, *Captain Planet,* to encourage young activists.

Sources and Further Reading

Bibb, Porter. *Ted Turner: It Ain't as Easy as It Looks: A Biography.* Boulder, CO: Johnson Books, 1997.

Lowe, Janet. *Ted Turner Speaks: Insight from the World's Greatest Maverick.* New York: Wiley, 1999.

Schonfeld, Reese. *Me and Ted against the World: The Unauthorized Story of the Founding of CNN.* New York: Cliff Street, 2001.

"Ted Turner: Prince of the Global Village." http://www.time.com/time/special/moy/1991.html. Accessed 2 November 2001.

Paul Watson (1950–)

Canadian Paul Watson became an environmental activist as a child after he befriended a beaver near his family's home in New Brunswick. Trappers killed the beaver, and an angry Watson enlisted his five siblings to help him destroy the trap lines. Leaving home at age fifteen after the death of his mother in childbirth, he joined the crew of a Norwegian freighter—a job that took him all over the world and provided him with valuable nautical knowledge.

From 1968 to 1974, Watson took courses in communications and media at Simon Fraser University, although he never earned a degree. But he did become a campus activist, volunteering with the American Indian Movement as a medic at Wounded Knee and participating in Greenpeace activities as one of its founding members in Vancouver, British Columbia, in 1972.

In 1971, he sailed into a nuclear test area in Alaskan waters and in 1975 gained media attention for placing his body between a harpoon and a whale. In 1977, Watson had a philosophical split from other Greenpeace members, notably Patrick Moore, another founder. Because Watson believed in physical engagement, he left the group and founded the Sea Shepherd Conservation Society in 1977. He has referred to Greenpeace as the "Avon ladies of the environmental movement" (Goldberg, 2001). Greenpeace, in turn, has labeled Sea Shepherd as a terrorist group.

After purchasing a ship in Britain and converting the trawler into a "conservation enforcement vessel," Watson developed a reputation for aggressive confrontation and use of the media to publicize environmental issues. He has been referred to as an ecopirate, a term he appears to relish. But he is also a leader and organizer. The Sea Shepherd Society had 45,000 members in 2001 with a $1.8 million budget (Personal communication to author, 2001), and Watson remains the guiding force behind the organization. He conducts many public-speaking events each year on the topic of marine conservation and has also held a part-time teaching position with the Pasadena College of Design, where he lectures on the history of environmental activism and ecology. His life has been profiled by a number of authors, and he has written several books about his exploits. A Montreal film production company purchased the rights to film his story in *Ocean Warrior.*

The group's tactics have resulted in Watson's facing three life sentences for a 1994 incident involving a Cuban fishing vessel off the Canadian coast. The jury found Watson not guilty of all charges of criminal mischief but convicted him of minor mischief, which carried a fine of $35 and a thirty-day jail sentence. He was imprisoned in Holland in 1997 and was convicted, in absentia, by the Norwegian government for his participation in sinking a Norwegian whaling boat in 1992 and ramming a Norwegian coast guard ship in 1994. He retains his title as captain of all the society's ships and confronts his critics by noting, "There are only two types of organizations—those that do and those that do mail-outs" (Watson, 1991).

Sources and Further Reading

"Captain Paul Watson: Founder and President." http://www. seashepherd.com/aboutus/people/paulwatson.html. Accessed 14 August 2001.

Goldberg., Kim. "Ocean Warrior." *Canadian Dimension* 35, no. 3 (May–June 2001): 5.

Hunter, Robert. *Warriors of the Rainbow: A Chronicle of the Greenpeace Movement.* New York: Holt, Rinehart, and Winston, 1979.

Personal communication to author. E-mail, 23 October 2001.

Rubin, Charles T. *The Green Crusade: Rethinking the Roots of Radical Environmentalism.* New York: Free Press, 1994.

Watson, Paul. *Sea Shepherd: My Fight for Whales and Seals.* New York: W. W. Norton, 1982.

———. "Sea Shepherd Defense Policy." *Earth Island Journal* 6, no. 4 (Fall 1991): 41–42.

6

Reports, Documents, Cases, and Testimony

One of the best ways to understand how interest groups work is to read some of the materials that they produce as well as to analyze court opinions, articles, and reports or other documents. This chapter provides a sampling of these materials that explain the tactics of the groups, including green party movements; explores their missions and goals; and expresses their perspectives on various environmental issues. In each selection, a brief paragraph provides the context for the document and its importance in understanding environmental activism.

Organization Reports and Documents

Major Conclusions and Recommendations, Commission for Racial Justice, United Church of Christ, 1987

In 1982, the United Church of Christ began investigating and challenging the number of communities dealing with toxic waste in residential areas. In 1986, the church's Commission for Racial Justice initiated two studies that focused on the extent to which African Americans, Hispanic Americans, Asian Americans, Pacific Islanders, Native Americans, and others were being exposed to hazardous waste in their communities. The commission was the first activist organization to study the topic, and the

173

results became the cornerstone of what was to become the environmental justice movement in America.

The findings of the analytical study on the location of commercial hazardous waste facilities suggest the existence of clear patterns which show that communities with greater minority percentages of the population are more likely to be the sites of such facilities. The possibility that these patterns happened by chance is virtually impossible, strongly suggesting that some underlying factor or factors, which are related to race, played a role in the location of hazardous waste facilities. Therefore, the Commission for Racial Justice concludes that, indeed race has been a factor in the location of hazardous waste facilities in the United States.

The findings of the descriptive study on the location of uncontrolled toxic waste sites suggest an inordinate concentration of such sites in Black and Hispanic communities, particularly in urban areas. This situation reveals that the issue of race is an important factor in describing the problem of uncontrolled toxic waste sites. We, therefore, conclude that the cleanup of uncontrolled toxic waste sites in Black and Hispanic communities in the United States should be given the highest possible priority.

These findings expose a serious void in present government programs addressing racial and ethnic concerns in this area. The report, therefore, strongly urges the formation of necessary offices and task forces by federal, state and local governments to fill this void. Among the many recommendations of this report, we shall call attention to the following:

We urge the President of the United States to issue an executive order mandating federal agencies to consider the impact of current policies and regulations on racial and ethnic communities.

We urge the formation of an Office of Hazardous Wastes and Racial and Ethnic Affairs by the United States Environmental Protection Agency. This office should insure that racial and ethnic concerns regarding hazardous wastes, such as the cleanup of uncontrolled sites, are adequately addressed. In addition, we urge the EPA to establish a National Advisory Council on Racial and Ethnic Concerns.

We urge state governments to evaluate and make appropriate revisions in their criteria for the siting of new hazardous waste facilities to adequately take into account the racial and socio-economic characteristics of potential host communities.

We urge the U.S. Conference of Mayors, the National Conference of Black Mayors and the National League of Cities to convene a national conference to address these issues from a municipal perspective.

We urge civil rights and political organizations to gear up voter registration campaigns as a means to further empower racial and ethnic communities to effectively respond to hazardous waste issues and to place hazardous waste issues at the top of the state and national legislative agendas.

We urge local communities to initiate education and action programs around racial and ethnic concerns regarding hazardous wastes.

We also call for a series of additional actions. Of paramount importance are further epidemiological and demographic research and the provision of information on hazardous wastes to racial and ethnic communities.

This report firmly concludes that hazardous wastes in Black, Hispanic, and other racial and ethnic communities should be made a priority issue at all levels of government. This issue is not currently at the forefront of the nation's attention. Therefore, concerned citizens and policy-makers, who are cognizant of this growing national problem, must make this a priority concern.

Source: *Executive Summary, Toxic Wastes and Race in the United States.* Commission for Racial Justice, United Church of Christ, 1987, 11–13.

The Valdez (CERES) Principles

After the oil spill caused by the tanker Exxon Valdez, *the nonprofit Coalition for Environmental Responsible Economies (CERES) started a public campaign to persuade corporations to take more responsibility for environmental protection. CERES is a membership organization made up of citizens, environmental professionals and organizations, public interest groups, and religious organizations. In 1989, the group released the Valdez Principles (now the CERES Principles) as a code of environmental conduct for corporate and business planning.*

By adopting these principles, we publicly affirm our belief that corporations have a responsibility for the environment, and must conduct all aspects of their business as responsible stewards of the environment by operating in a manner that protects the Earth. We believe that corporations must not compromise the ability of future generations to sustain themselves. We will update our practices constantly in light of advances in technology and new understandings in health and environmental science. In collaboration with CERES, we will promote a dynamic process to ensure that the Principles are interpreted in a way that accommodates changing technology and environmental realities. We intend to make consistent, measurable progress in implementing these Principles and to apply them to all aspects of our operations throughout the world.

1. Protection of the Biosphere: We will reduce and make continual progress toward eliminating the release of any substance that may

cause environmental damage to the air, water, or the earth or its inhabitants. We will safeguard all habitats affected by our operations and will protect open spaces and wilderness, while preserving biodiversity.

2. Sustainable Use of Natural Resources: We will make sustainable use of renewable natural resources, such as water, soils and forests. We will conserve nonrenewable natural resources through efficient use and careful planning.

3. Reduction and Disposal of Wastes: We will reduce and where possible eliminate waste through source reduction and recycling. All waste will be handled and disposed of through safe and responsible methods.

4. Energy Conservation: We will conserve energy and improve the energy efficiency of our internal operations and of the goods and services we sell. We will make every effort to use environmentally safe and sustainable energy sources.

5. Risk Reduction: We will strive to minimize the environmental, health and safety risks to our employees and the communities in which we operate through safe technologies, facilities and operating procedures and by being prepared for emergencies.

6. Safe Products and Services: We will reduce and where possible eliminate the use, manufacture or sale of products and services that cause environmental damage or health or safety hazards. We will inform our customers of the environmental impacts of our products or services and try to correct unsafe use.

7. Environmental Restoration: We will promptly and responsibly correct conditions we have caused that endanger health, safety or the environment. To the extent feasible, we will redress injuries we have caused to persons or damage we have caused to the environment and will restore the environment.

8. Informing the Public: We will inform, in a timely manner, everyone who may be affected by conditions caused by our company that might endanger health, safety or the environment. We will regularly seek advice and counsel through dialogue with persons in communities near our facilities. We will not take any action against employees for reporting dangerous incidents or conditions to management or appropriate authorities.

9. Management Commitment: We will implement these Principles and sustain a process that ensures that the Board of Directors and Chief Executive Officer are fully informed about pertinent environmental issues and are fully responsible for environmental policy. In selecting our Board of Directors, we will consider demonstrated environmental commitment as a factor.

10. Audits and Reports: We will conduct an annual self-evaluation of our progress in implementing these Principles. We will support the timely creation of generally accepted environmental audit procedures.

We will annually complete the CERES Report, which will be made available to the public.

Source: Coalition for Environmentally Responsible Economies. http://www.ceres.org. Accessed 7 February 2002.

The Twenty-Five Goals of the Wise Use Movement

In August 1988, the Center for the Defense of Free Enterprise held a conference in Reno, Nevada, that drew nearly 300 U.S. and Canadian leaders in natural resources issues. The Multiple Use Strategy Conference was attended by executives of public interest membership organizations, unaffiliated citizens with "a common passion for life on this planet," trade groups and government representatives, and those representing such extractive industries as timber, mining, and petroleum.

The subject of the conference debate was the relationship between humans and their environment, with a focus on the development of natural resources, the role of the environmental movement, and restrictions on the use of the earth's ecosystem. The resulting document, The Wise Use Agenda, *represents the conference's recommendations and serves as the platform for the contemporary wise use movement in the United States. The document was transmitted to President George H. W. Bush with the top twenty-five goals, which are listed below.*

1. Initiation of a wise use public education project
2. Immediate wise development of the petroleum resources of the Arctic National Wildlife Refuge
3. The Inholder Protection Act
4. Passage of the Global Warming Protection Act
5. Creation of the Tongass National Forest Timber Harvest Area
6. Creation of a National Mining System
7. Passage of the Beneficial Water Rights Act
8. Commemorate the 100th anniversary of the founding of the Forest Reserves by William Steele Holman
9. The Rural Community Stability Act
10. Creation of a National Timber Harvest System
11. National Parks Reform Act
12. Pre-Patent Protection of Pest Control Chemicals
13. Create the National Rangeland Grazing System
14. Compassionate Wilderness Policy
15. National Industrial Policy Act
16. Truth in Regulation Act

17. Property Rights Protection
18. Endangered Species Act Amendments
19. Obstructionism Liability
20. Private Rights in Federal Lands Act
21. Global Resources Wise Use Act
22. Perfect the Wilderness Act
23. Standing to Sue in Defense of Industry
24. National Recreation Trails Trust Fund
25. The End of the "Let Burn" Policy

Source: Gottlieb, Alan, ed. *The Wise Use Agenda: The Citizen's Policy Guide to Environmental Resource Issues.* Bellevue, WA: Free Enterprise Press, 1989, 5–18.

Developing and Implementing the Dunn Land Ethic

Dunn, Wisconsin, is a small community south of the state's capital of Madison. In 1972, the town's citizens decided to throw out the existing town board to elect individuals who were more responsive to their concerns on increasingly complex land use issues. After developing an inventory of the town's geography and natural resources, from bedrock geology and lakes and ponds to historical sites and prairies, "We came to know our place. We liked what we found. We also decided to care for it and keep it" (DeWitt, 1996).

What emerged was the Town of Dunn Open Space Preservation Handbook—a land stewardship plan—that became legally titled as the Town of Dunn Subdivision Ordinance. In 1995, the town received the Renew America Award for exemplary growth management. The process by which the town's citizens took control of land use is described by one of its leaders, who calls the results of their activism a stewardship ethic and stewardship plan.

1. Observing and Evaluating (Problem Recognition)

The very first step was to observe and evaluate what was happening around us. Everywhere we looked we found that when farmers reached retirement age that their land was sold to people who would subdivide the land and build houses on it. There was no "rhyme or reason" to it. We called it "helter-skelter" development. Natural areas were being threatened, altered, and destroyed. The practice of farming was declining. "Development" meant only residential subdivisions to the neglect of the rich meaning of this word. The likely future was a landscape of houses and shopping malls with abandoned farms and habitats that stood in waiting for the time they would become houses and shopping areas too. Life was being transformed from one that had visible connections with creation to one where these connections were broken. People were becoming enmeshed with the products of human activity—fast travel on expressways, security and protection systems, maintaining lawns as monocultures, immersion in virtual worlds in other galaxies, anticipation of escape vacations. And all of this at the expense of disconnection from the sustaining land and its remarkable life. People were becoming aliens to their own place.

2. Forming an Energetic and Committed Town Board and Constituency

People had to be found who were willing to change their goals in life toward knowing and serving their place with full dedication to maintaining and restoring human community, agriculture, natural systems, and the plants and animals that with their physical environments form the wetlands, woodlands, streams, and prairies. Basic to this was forming an energetic and committed constituency to engage in the necessary discussions, raise needed funds, serve on committees and commissions, organize festive and educational events, write brochures, lead field trips, engage children and adults, and work to put the right people into town office. The constituency must be very substantial. In the Town of Dunn in the early years of the effort to build and implement a land ethic, this need was met by generating such interest that approximately ten people were able and willing to run for town office, 100 were able and willing to do substantial work for the town within 24 hours, and that 1,000 were able and willing to do something substantial for the town with a week's notice.

3. Putting Initial Control Measures into Place

Getting into political office can be very disappointing if one expects really to be able to change things. This is due to the fact that the power to do things often resides mainly outside of the government unit within which you are operating. In the Town of Dunn, much of the power was vested in the State of Wisconsin and in Dane County. This "outside" power, for example, includes the State Highway Department, the State Department of Natural Resources, the County Board, the County Highway Department, etc. And so it is necessary to do a careful study of who has the power to do the things that need to be done.

The Town of Dunn has to discover where it currently had power to control its destiny of the town. Where it had such power, as the Town of Dunn has over its road system, it had to be exercised toward fulfillment of ecological and social integrity. Thus, for awhile, "Road Access Permits" were employed by the town to provide foreknowledge to the town of projects approved by the County that required access off of town roads. We used our power over roads to make it necessary for people first to come to the town before they got a permit from the county for changing a land use. Where the town did not have power, it had to find routes to achieving such power. This required careful selection of a bright lawyer who persistently sought ways for achieving goals, rather than consistently told the town what it could not do. While means of achieving necessary power were being sought, discussion, reason, and persuasion had to be employed diligently.

4. Implementing a Moratorium

A moratorium was declared on all land divisions in the Town for two years. This provided the time needed to think about who we are, what we had been and where we were headed. And the moratorium allowed us to put into place whatever was needed to assure a future with social and ecological integrity. For us in the Town of Dunn, this did not give immediate quiet because it resulted in a whole series of law suits filed against us during the first few months. These, however, were maneuvers designed to intimidate us. Again, a good lawyer for the town gave us the insight and encouragement we needed to persist in our work. To help us handle legal challenges we built up a fund of $100,000 by not building roads for one year. This allowed us to stand firm in the legal challenges that were sometimes hurled in at us.

5. Executing a Land and Cultural Inventory

Our inventory was extensive, covering the various ecosystems, biodiversity, agriculture, and human community past and present. We made good use of the expertise available from the university and from the citizens of the town in getting this put together. Some of the maps were made by citizens in a local drafting shop after hours, at no cost to the town. We discovered, as anyone would if they explored their own expertise, that there were many talented people among us that could do the work. Results of the inventory were published in the Town of Dunn Open Space Handbook in 1979, and an accompanying data book.

6. Adopting the Land Stewardship Plan

Everything we had learned from doing the inventory, and from reasoning with each other in hundreds of meetings (we met weekly or more for over three years and more), was put together into a plan for the town. It was entitled, the Town of Dunn Land Use Plan. While the board has the authority to adopt this plan, we chose a referendum instead. This got everyone engaged in the process. We gave advance notice of the upcoming referendum on the plan many months in advance and found that this was ideal for engaging people in discussion, deliberation, and self-education about their town and its future.

7. Codifying the Stewardship into Ordinances

Based upon these documents a Town of Dunn Subdivision Ordinance was developed and adopted, supported by the implementation of a new Agricultural Conservancy land use zoning.

8. Enforcing the Ordinance Consistently and Uniformly

These have been uniformly enforced ever since through their administration by the Town of Dunn Plan Commission and the Town of Dunn Board. Plans and ordinances without enforcement are worthless. And non-uniform or inconsistent enforcement sets bad precedents that will destroy what has been accomplished. Provisions had been made to change the plan through a highly deliberative procedure to address problems and situations not foreseen at its original writing. Problems were addressed not by making exceptions, but by modifying the plan through thoughtful and deliberative work.

9. Nurturing the Stewardship Ethic That Is Expressed in the Ordinance

What was codified into ordinances and written into the plan was not enough; it reflected the underlying land ethic but was not that ethic itself. This means that the land ethic must be nurtured continuously. In Dunn this is accomplished through an official newspaper produced by the town, annual festive and educational events, and deliberative development of new policies that further the purposes of the Dunn Land Ethic. Current work to raise property taxes in order to develop a fund for the purchase of development rights is an example.

10. Publishing the Stewardship Ethic in Land and Life

Finally, the real product of the whole process is the life of the people of Dunn and the landscape of Dunn. Unless the Dunn Land Ethic is published in land and life it is not a real ethic, but something merely recorded in books and documents. The measure of the success of this ethic is what one finds in human lives and in the landscape of the town. To the extent this improves or regresses, adjustments may need to be made. The proof of the Dunn Land Ethic is its positive expression in land and life.

Source: DeWitt, Calvin B., Director, Au Sable Institute. "Community Mobilization: A Case Study of the Town of Dunn." October 1996.

Legal Petition to the Food and Drug Administration on Genetically Engineered Foods

On 21 March 2000, a coalition of scientific, consumer, environmental, and farm organizations, led by the Center for Food Safety, filed a legal petition with the federal Food and Drug Administration (FDA) demanding the development of a thorough premarket and environmental testing regime for genetically engineered foods and the mandatory labeling of genetically engineered foods. In this selection from the executive summary to the petition, the groups seek to change the way in which the regulatory process is conducted.

The legal petition is an attempt to change this arbitrary and dangerous government policy and to ensure that genetically engineered foods are subjected to a mandatory pre-market testing and labeling regulatory regime. The action demands that the FDA alter existing regulations to address the many potential human health and environmental impacts of genetically engineered foods. Petitioners seek the removal of all genetically engineered foods from the marketplace unless and until the FDA takes the following actions:

• Rescind its 1992 Policy on genetically engineered foods and implement new regulations that comply with the legal requirements of the Federal Food, Drug and Cosmetic Act and require all genetically engineered foods be subject to the pre-market safety assessment procedures embodied in the food additive petition process. This would trigger the FDA's need for specific safety testing and approval of each genetically engineered food before it could be considered "generally recognized as safe";

• Enact additional regulatory requirements within the food additive petition review process for genetically engineered foods and food additives that assess potential allergenicity, toxicity and unintended effects of such foods. The petition outlines a number of testing and analysis techniques that could make up part of this regulatory requirement;

• Complete a programmatic environmental impact statement assessing the FDA's program on genetically engineered foods, and ensure that each genetically engineered food subject to food addition petition review be subject to environmental impact analysis; and

• Mandate the labeling of all genetically engineered foods entering the marketplace. Petitioners assert that the current FDA policies on genetically engineered foods violate the Federal Food, Drug and Cosmetic Act because the law requires the mandatory labeling of genetically engineered foods because of the "material" changes made to such foods.

Source: "Executive Summary, Legal Petition to the Food and Drug Administration on Genetically Engineered Foods." http://www.centerforfoodsafety.org. Accessed 21 March 2000.

Meet the E.L.F.

Radical activists engaged in acts of environmental sabotage, or ecotage, such as those described in Chapter 1, often work alone but are rarely part of an organized group. They do not have lobbyists in Washington, D.C., do not work to elect candidates, and do not send out mass mailings or hold conferences. A typical example is the Earth Liberation Front (E.L.F.). The statement reproduced below provides some insight into the E.L.F. "mission" while making clear that individuals act on their own.

The Earth Liberation Front is an international underground movement consisting of autonomous groups of people who carry out direct action according to E.L.F. guidelines. Since 1997, E.L.F. cells have carried out dozens of actions resulting in over $30 million in damages.

Modeled after the Animal Liberation Front, the E.L.F. is structured in such a way as to maximize effectiveness. By operating in cells (small groups that consist of one to several people), the security of group members is maintained. Each cell is anonymous not only to the public but to one another, free to continue conducting actions.

As the E.L.F. structure is non-hierarchical, individuals involved control their own activities. There is no centralized organization or leadership tying the anonymous cells together. Likewise, there is no official "membership." Individuals who choose to do actions under the banner of E.L.F. are driven only by their personal conscience or decisions taken by their cell while adhering to the stated guidelines.

Who are the people carrying out these activities? Because involved individuals are anonymous, they could be anyone from any community. Parents, teachers, church volunteers, your neighbor, or even your partner could be involved. The exploitation and destruction of the environment affects all of us—some people enough to take direct action in defense of the earth.

Any direct action to halt the destruction of the environment and adhering to the strict nonviolence guidelines, listed below, can be considered an E.L.F. action. Economic sabotage and property destruction fall within these guidelines.

Earth Liberation Front Guidelines:

*   To inflict economic damage on those profiting from the destruction and exploitation of the environment.
*   To reveal and educate the public on the atrocities committed against the earth and all species that populate it.
*   To take all necessary precautions against harming any animal, human and non-human.

There is no way to contact the E.L.F. in your area. It is up to each committed person to take responsibility for stopping the exploitation of the natural world. No longer can it be assumed that someone else is going to do it. If not you who, if not now when?

Source: http://www.earthliberationfront.org.

Green Party Documents

The green party movement represents an extension of environmental activism, although the formation of parties and their successes vary considerably from one nation to another. In some European nations, for example, the Greens have been able to utilize political systems that use proportional representation to their advantage, gaining seats within legislative bodies. In the United States, the victories have been almost exclusively at the local level.

Over time, the philosophical basis of the Greens has gone through considerable change. Although usually identified as a movement seeking environmental sustainability, these three documents show how the green party has expanded its vision and mission to a much broader range of issues. Once built on ten key values, the national and international Green movement is now based on a lengthy list of actions and policy preferences.

Ten Key Values of the Green Committees of Correspondence

In 1984, the Committees of Correspondence represented the fledgling Green movement in the United States. In 1989, the name was changed to Green Committees of Correspondence, and the following ten key values were adopted. Each value included a series of questions related to the topic and formed the basis for the group's activism. In 1991, the name was changed to the Greens/Green Party USA—a confederation of local and state affiliates.

1. Ecological Wisdom
2. Grassroots Democracy
3. Social Justice
4. Nonviolence
5. Decentralization
6. Community-Based Economics
7. Feminism
8. Respect for Diversity
9. Global Responsibility
10. Sustainability

Source: Global Greens. http://www.global.greens.org.

Platform Planks, the Green Party USA

At its annual Green Congress, held 26–28 May 2000 in Chicago, the Greens/Green Party USA adopted a platform that, although not binding on state and local affiliates, represents the immediate policy goals of the party. In the Preamble to the Platform, the delegates noted, "To the social movements, the Greens say that in order to progress toward a democratic society, we must resolve the ecological crisis so that people are still around to enjoy democracy. To the environmental movements, Greens say that in order to have an ecological society, we must have a democratic society so that people have the power to choose ecological sustainability. To survive, we must have ecological sustainability. To choose ecological sustainability, we must have the power of democracy."

1. An Economic Bill of Rights
   - Universal Social Security
   - Jobs for All
   - Living Wages
   - 30-hour Work Week
   - Social Dividends
   - Universal Health Care
   - Free Child Care
   - Lifelong Public Education
   - Affordable Housing
2. Grassroots Democracy
   - Community Assemblies
   - A Proportional, Single-Chamber US Congress
   - Environmental Home Rule
   - Average Workers' Pay for Elected Officials
   - DC Statehood

3. Fair Elections
   - Proportional Representation
   - Preference Voting
   - Public Campaigns and Party Financing
   - Fair Ballot Access
   - Eliminate Mandatory Primaries
4. Ecological Conversion
   - Ecological Production
   - Renewable Energy
   - Biotechnology—No Patents on Life; No Transgenic Organisms
   - Environmental Defense and Restoration
   - Environmental Justice
   - A Just Transition
5. Sustainable Agriculture
   - Fair Farm Price Supports
   - Subsidize Transition to Organic Agriculture
   - Support Small Farmers
   - Break Up Corporate Agribusiness
6. Economic Democracy
   - Eliminate Corporate Personhood
   - End Limited Corporate Liability
   - Federal Chartering of Interstate Corporations
   - Periodic Review of Corporate Charters
   - Strengthen Anti-Trust Enforcement
   - Democratic Production
   - Workplace Democracy
   - Worker Control of Worker Assets—Pension Funds and ESOP Shares
   - Democratic Conversion of Big Business
   - Democratic Conversion of Small and Medium Business
   - Democratic Banking
   - Democratize Monetary Policy and the Federal Reserve System
7. Progressive and Ecological Taxes
   - Ecological Taxes
   - Simple, Progressive Income Taxes
   - Eliminate Regressive Payroll Taxes
   - Guaranteed Adequate Income
   - Maximum Income
   - End Corporate Welfare
   - Wealth Tax
   - Inheritance Tax
   - Stock and Bond Transfer Tax
   - Currency Speculation Tax
   - Advertising Tax

- Federal Revenue Sharing
- Ecological and Feminist Accounting
8. Human Rights and Social Justice
    - End Institutionalized Racism, Sexism, and Oppression of People with Disabilities
    - African American Reparations
    - Indian Treaty Rights
    - Immigrant Rights
    - Reproductive Freedom
    - Comparable Worth
    - End Discrimination Against Lesbian, Gay, Bisexual, and Transgendered People
    - Same-Sex Marriage
9. Criminal and Civil Justice Reforms
    - Abolish the Death Penalty
    - Prosecute Police Brutality
    - End Political and Racial Persecution by the Criminal Justice System
    - Restorative Justice
    - Legal Aid
    - Fight Corporate Crime
    - Oppose Tort Reform That Limits Class Action Lawsuits and Caps Victims' Compensation
    - Civil Liberties
    - End the "War on Drugs"
10. Labor Law Reforms
    - Repeal Repressive Labor Laws
    - A Workers' Bill of Rights
    - Expand Workers' Rights to Organize and Enjoy Free Time
11. Revitalize Public Education
    - Equalize School Funding with Federal Revenue Sharing
    - Decentralized Administration
    - Class Size Reduction
    - Preschool Programs
    - After School Programs
    - Children's Health
    - Improve Teacher Training and Pay
    - Multicultural Teaching Staffs
    - Tuition-Free Higher Education
    - Oppose the Privatization of Public Schools
    - Curriculum for a Multicultural Participatory Democracy
    - Support Bilingual Education
12. Free, Diverse and Uncensored Media
    - Infodiversity
    - Support Nonprofit and Noncommercial Media

* Real Public Broadcasting
* Regulate Public Airwaves in the Public Interest
* Antitrust Actions to Break Up Media Conglomerates
13. International Solidarity
   * A Global Green Deal
   * Peace Conversion
   * Peace Dividend
   * Unilateral Nuclear, Biological, and Chemical Disarmament
   * Cooperative Security
   * Democratize the United Nations
   * A Pro-Democracy Foreign Policy
   * End Global Financial Exploitation
   * Fair Trade

Source: The Greens/Green Party USA. http://www.greenparty.org. Accessed 2 March 2002.

Preamble, Charter of the Global Greens

In Canberra, Australia, in 2001, the Global Greens "defined what it means to be Green in the new millennium" ("Charter of the Global Greens," 2002). This group serves as the international network of Green parties and political movements, with its own principles that guide policy actions. The Preamble, which represents the collective statement of the Global Greens, outlines the beliefs upon which the principles are identified. The document contrasts sharply with the Platform of the Greens/Green Party USA, especially in its attention to environmental issues.

We, as citizens of the planet and members of the Global Greens,
United in our awareness that we depend on the Earth's vitality, diversity and beauty, and that it is our responsibility to pass them on, undiminished or even improved, to the next generation
Recognizing that the dominant patterns of human production and consumption, based on the dogma of economic growth at any cost and the excessive and wasteful use of natural resources without considering Earth's carrying capacity, are causing extreme deterioration in the environment and a massive extinction of species
Acknowledging that injustice, racism, poverty, ignorance, corruption, crime and violence, armed conflict and the search for maximum short term profit are causing widespread human suffering
Accepting that developed countries through their pursuit of economic and political goals have contributed to the degradation of the environment and of human dignity
Understanding that many of the world's peoples and nations have been impoverished by the long centuries of colonization and

exploitation, creating an ecological debt owed by the rich nations to those that have been impoverished

Committed to closing the gap between rich and poor and building a citizenship based on equal rights for all individuals in all spheres of social, economic, political and cultural life

Recognizing that without equality between men and women, no real democracy can be achieved

Concerned for the dignity of humanity and the value of cultural heritage

Recognizing the rights of indigenous people and their contribution to the common heritage, as well as the right of all minorities and oppressed peoples to their culture, religion, economic and cultural life

Convinced that cooperation rather than competition is a prerequisite for ensuring the guarantee of such human rights as nutritious food, comfortable shelter, health, education, fair labor, free speech, clean air, potable water and an unspoilt natural environment

Recognizing that the environment ignores borders between countries and

Building on the Declaration of the Global Gathering of Greens at Rio in 1992

Assert the need for fundamental changes in people's attitudes, values, and ways of producing and living

Declare that the new millennium provides a defining point to begin that transformation

Resolve to promote a comprehensive concept of sustainability which

protects and restores the integrity of the Earth's ecosystems, with special concern for biodiversity and the natural processes that sustain life

acknowledges the interrelatedness of all ecological, social and economic processes

- balances individual interests with the common good
- harmonizes freedom with responsibility
- welcomes diversity within unity
- reconciles short term objectives with long term goals
- ensures that future generations have the same right as the present generation to natural and cultural benefits

Affirm our responsibility to one another, to the greater community of life, and to future generations

Commit ourselves as Green parties and political movements from around the world to implement these interrelated principles and to create a global partnership in support of their fulfillment.

Source: "Charter of the Global Greens." http://www.global.greens.org. Accessed 2 March 2002.

Government Documents

Another way to look at the role of environmental activism is to examine some of the documents that have been produced by the U.S. government and by state and local entities. There are millions of reports, statements, press releases, agreements, and other materials dealing with environmental issues and the people they affect. This sampling shows the variations both in the types of problems addressed and the audiences to whom the documents are directed.

Principles: American Indian Tribal Rights, Federal-Tribal Trust Responsibilities, and the Endangered Species Act

Native American tribal governments have often been in conflict with federal environmental legislation, especially when they are asked to help implement and enforce federal statutes. Congress may, for example, be unclear in its intent or definitions within a law, leaving it open to interpretation and confusion. In this instance, the secretary of the interior and the secretary of commerce, both of whom have authority in the implementation of the Endangered Species Act of 1973, are attempting to clarify their relationships with tribal groups with whom the government has trust responsibility and treaty obligations. At issue is the conflict between the agencies' mission of protecting endangered species and the rights of native peoples to tribal sovereignty over their lands, including the use of sacred plants and animals. This secretarial order comes in response to the concerns of tribal groups who perceived that the government was not cooperative and underscores the secretaries' intent to take into consideration the impacts of agency actions and policies when dealing with Indian culture, religion, and spirituality. It is an example of how formal agreements are often needed to gain cooperation between groups and the government when sensitive environmental issues are at stake.

• Principle 1: The Departments shall work directly with Indian tribes on a government-to-government basis to promote healthy ecosystems.
• Principle 2: The Departments shall recognize that Indian lands are not subject to the same controls as federal public lands.

• Principle 3: The Departments shall assist Indian tribes in developing and expanding tribal programs so that healthy ecosystems are promoted and conservation restrictions are unnecessary.

A. The Departments shall take affirmative steps to assist Indian tribes in developing and expanding tribal programs that promote healthy ecosystems.

B. The Departments shall recognize that Indian tribes are appropriate governmental entities to manage their lands and tribal trust resources.

C. The Department, as trustees, shall support tribal measures that preclude the need for conservation restrictions.

• Principle 4: The Departments shall be sensitive to Indian Culture, Religion, and Spirituality.

• Principle 5: The Departments shall make available to Indian tribes information related to tribal trust resources and Indian lands, and to facilitate the mutual exchange of information, shall strive to protect sensitive tribal information from disclosure.

Source: Secretarial Order #3206, Secretary of the Interior/Secretary of Commerce, 5 June 1997.

Statement of Ron Arnold before the House Subcommittee on Crime Hearing on Acts of Terrorism by Radical Environmental Organizations

During the second session of the 105th Congress in 1998, the House of Representatives Subcommittee on Crime, chaired by Representative Bill McCollum of Florida, convened a hearing to consider the problems associated with violent acts by radical environmental organizations, or "ecoterrorism." Among the witnesses at the hearing was Ron Arnold, a leader in the wise use movement of environmental opposition groups. In this prepared statement, he calls upon Congress to enact an amendment to a 1993 statute that would provide legal protection and recourse to those harmed by ecoterror crimes.

Mr. Chairman and Members of the Committee, my name is Ron Arnold. I am testifying as the executive vice president of the Center for the Defense of Free Enterprise, a nonprofit citizen organization based in Bellevue, Washington. The Center has approximately 10,000 members nationwide, most of them in rural natural resource industries.

Mr. Chairman, the Center does not accept government grants and is in full compliance with House Rule XI, clause 2(g). Mr. Chairman, I would like to thank you on behalf of our members for holding this hearing today. It is long overdue. For the past five years our members have routinely contacted our headquarters to report crimes committed against them of a type we have come to call ecoterrorism, that is, a crime committed to save nature. These crimes generally take the form of equipment vandalism, but may include package bombs, blockades using physical force to obstruct workers from going where they have a right to go, and invasions of private or government offices to commit the crime of civil disobedience. So you can see, Mr. Chairman, the range of ecoterror crimes ranges from the most violent felonies of attempted murder to misdemeanor offenses such as criminal trespass. But they are all crimes. I am not here to discuss noncriminal actions that do not result in arrests and convictions.

My organization's membership is nationwide. There is no region of the United States where I have not received complaints from members about being victimized by ecoterrorists. It is a broad and pervasive crime that is seriously under-reported because the victimized are terrorized and fear reprisals, copycat crimes, or in the case of corporations, loss of customer confidence and resulting drops in share prices.

I am the author of a book on the subject of this hearing, titled EcoTerror. In this book I have reported the tactics of organized vandalism called by environmentalists "monkeywrenching," which means sabotage against goods producers and their equipment in order to save nature. Ecoterrorism has been studied by social scientists with illuminating results. In particular, the tactics of a group known as Earth First! have been described in the Academy of Management Journal in a study entitled Acquired Organizational Legitimacy Through Illegitimate Actions. I request that pages 699, 715, 716, and 717 of this study be made a part of the record.

I interviewed the lead author of this study to verify its contents. Kimberly Elsbach told me that the data were gathered directly from Earth Firsters who allowed her to witness criminal acts on condition that she destroy her notes as soon as her scholarship no longer needed them. One of the most pertinent tactics she discovered was called "decoupling," which is a set of techniques denying the crime while deploring the conditions that caused the perpetrators to become so frustrated they committed the crime. Thus decoupling throws blame for the crime on the victim while it denies guilt. However, law enforcement officers have concluded that in fact Earth Firsters were the perpetrators, a conclusion drawn as a result of several arrests and convictions in which the defendant admitted connection to Earth First.

As Earth First in recent years has tried to mainstream itself, ecoterror crimes have become more destructive to their wishes for a good public image. Therefore, Judi Bari, an Earth First Leader, wrote an article in the Earth First Journal recommending that a decoupling group call itself Earth Liberation Front in order to create deniability for Earth Firsters crimes. I document this in my book EcoTerror on page 270, which I respectfully request be made part of the record. In fact, the Earth Liberation Front has subsequently become a well-known entity to law enforcement. Furthermore, Mr. Chairman, the Earth Liberation Front and the Animal Liberation Front signed a joint communique stating their solidarity and blending. I have been able to determine that certain criminal Earth Firsters, Earth Liberation Front members and the Animal Liberation Front members are the same people. Examples are David Barbarash and Darren Thurston, convicted felons now under indictment in Canada for attempted murder by pipe bombs, were at one time Earth Firsters. I am stating that there is no difference between ecoterrorism and animal rights terrorism. The perpetrators are in large part the same, and the solidarity of action is openly declared.

These crimes to save nature are difficult to solve for law enforcement. The solution is to extend federal protection to loggers, miners, fishermen, farmers and ranchers, and others who are the most frequent targets of ecoterrorist attack. A simple way to accomplish this would be to add those classes of people to the list of persons protected by the Animal Enterprise Protection Act of 1993. That law federalized crimes of property damage over $10,000 or that resulted in dismemberment or death to a human being as a result of attacks on animal enterprises. A simple amendment would create the Resource Enterprise Protection Amendment of 1998 by adding to the list of protected persons loggers, miners, fishermen, farmers, trappers, ranchers, food outlets and processors and all resource enterprises subject to ecoterror crimes. This law also needs a citizens attorneys general clause to allow harmed parties to seek relief in federal court, and it needs a periodic report to Congress. The existing Animal Enterprise Protection Act also needs to be reviewed because its enforcement has proven to be lax and virtually ineffectual. Congressional oversight of its enforcement is badly needed.

I feel that this modest proposal would meet with congressional approval and would go far to protecting the interests of all natural resource producers in America.

Thank you again, Mr. Chairman, for holding this hearing.

Source: U.S. House Committee on the Judiciary, Subcommittee on Crime. *Acts of Ecoterrorism by Radical Environmental Organizations: Hearing before the Subcommittee on Crime of the Committee on the Judiciary.* 105th Cong., 2d sess., 9 June 1998.

Legal Cases

As was mentioned earlier in this book, litigation has become one of the most common strategies used by environmental activist groups. Sometimes the courts are used as a way of seeking enforcement for a statute. In other situations, groups seek redress when they believe that the government has done something wrong or unconstitutional. During the early years of the environmental movement, numerous cases were brought by citizen groups and by public interest law firms against government officials. These selections show the evolution of the courts' rulings as the door to litigants is gradually opened.

Citizens to Preserve Overton Park v. Volpe (1971)

One of the provisions of the Administrative Procedure Act allows that the actions of the government and federal agencies are subject to review by the courts, except where statutes specifically preclude judicial review or when an agency has discretion by law. Many of the reviews begin as lawsuits brought by environment groups, as in the following case. Activists and conservation groups in Memphis, Tennessee, challenged the authorization of expenditures to build a six-lane interstate highway through Overton Park, a 342-acre city park near the city's center. Within the park are a zoo, municipal golf course, outdoor theater, nature trails, picnic areas, bridle path, and 170 acres of forest. The proposed highway would have separated the zoo from the rest of Overton Park and would also have destroyed 26 acres of parkland.

A citizens group brought suit against Secretary of Transportation John A. Volpe, who had authorized the project. They contended that the Department of Transportation Act of 1966 and the Federal-Aid to Highway Act of 1968 prohibited the use of federal funds to finance construction through public parks if a "feasible and prudent" alternative route exists. Here, the Supreme Court examines reviewability and rules in favor of the Citizens to Preserve Overton Park.

Mr. Justice Marshall delivered the opinion of the Court: Congress clearly did not intend that cost and disruption of the community were to be ignored by the secretary. But the very existence of the statute indicates that protection of parkland was to be given paramount importance. The few green havens that are public parks were not to be lost unless there were truly unusual factors present in a particular case or the cost of community disruption resulting from alternative routes reached extraordinary magnitudes. If the statutes are to have any meaning, the Secretary cannot approve the destruction of parkland unless he finds that alternative routes present unique problems.

Source: *Citizens to Preserve Overton Park v. Volpe*, 401 U.S. 402 (1971).

Sierra Club v. Morton (1972)

In 1969, the U.S. Forest Service gave its approval to Walt Disney Enterprises for a planned $35 million recreational complex in California's Sierra Nevada. The proposed development would be built on eighty acres within the Mineral King valley, part of a game refuge within Sequoia National Park. The final Disney plan included motels, restaurants, swimming pools, parking lots, and other structures designed to accommodate 14,000 daily visitors. In addition, the state of California would build a twenty-mile access road to the resort, and ski lifts and trails would be constructed on adjacent mountain slopes. The Sierra Club brought suit against the secretary of the interior, Rogers Morton, attempting to stop the development from being built within the protected area. The group argued that the proposed development was contrary to federal laws and regulations related to the protection and preservation of national parks, forests, and game refuges.

The issue before the U.S. Supreme Court was whether the Sierra Club had standing—a concept that was tested in this case. In order to have standing, the petitioners must show that they have a personal interest in the outcome of the case and that they will suffer an actual or potential injury that can only be prevented by the court. If the person or group bringing the suit does not have standing, the court will not even hear arguments about the merits of the case. The lack of standing prevented many environmental groups from attempting to litigate cases because they were unable to show how they would be "injured."

This excerpt from the U.S. Supreme Court decision explains why standing is so important in determining the success or failure of environmental activists to bring an issue to the judicial arena. Here, the Court decides that the Sierra Club lacked standing in its suit because the organization failed to show that its members would be significantly affected by the Disney project.

Justice Stewart delivered the opinion of the Court: The injury alleged by the Sierra Club will be incurred entirely by reason of the change in the uses to which Mineral King will be put, and the attendant change in the aesthetics and ecology of the area. . . . The complaint alleged that the development "would destroy or otherwise adversely affect the scenery, natural or historic objects and wildlife of the park and would impair the enjoyment of the park for future generations." We do not question that that type of harm may amount to an "injury in fact" sufficient to lay the basis for standing under Section 10 of the Administrative Procedure Act. Aesthetic and environmental well-being, like economic well-being, are important ingredients of the quality of life in our society, and the fact that particular environmental interests are shared by the many rather than the few does not make them less deserving of legal protection through the judicial process. But the "injury in fact" test requires more than an injury to a cognizable interest. It requires that the party seeking redress be himself among the injured.

The impact of the proposed changes in the environment of Mineral King will not fall indiscriminately upon every citizen. The alleged injury will be felt only by those who use Mineral King and Sequoia National Park. . . . The Sierra Club failed to allege that it or its members would be affected in any of their activities or pastimes by the Disney development. Nowhere in the pleadings or affidavits did the Club state that its members use Mineral King for any purpose, much less that they use it in any way that would be significantly affected by the proposed actions of the respondents. . . .

But broadening the categories of injury that may be alleged in support of standing is a different matter from abandoning the requirement that the party seeking review must himself have suffered an injury. Some courts have indicated a willingness to take this latter step by conferring standing upon organizations that have demonstrated "an organizational interest in the problem" of environmental or consumer protection. It is clear that an organization whose members are injured may represent those members in a proceeding for judicial review. . . .

The requirement that a party seeking review must allege facts showing that he is himself adversely affected does not insulate executive actions from judicial review, nor does it prevent any public interests from being protected through the judicial process. It does serve . . . to put the decision as to whether review will be sought in the hands of those who have a direct stake in the outcome. That goal would be undermined were we to construe the APA to authorize judicial review at the behest of organizations or individuals who seek to do no more than vindicate their own value preferences through the judicial process.

Source: *Sierra Club v. Morton*, 405 U.S. 727 (1972).

The Smithfield Litigation

Between 1984 and 1997, the Natural Resources Defense Council (NRDC), a public interest environmental law firm, and one of its non-profit regional partners, the Chesapeake Bay Foundation (CBF), attempted to pressure the Virginia State Water Control Board to enforce provisions of the federal Clean Water Act. One of their targets was a consortium of companies generally referred to as Smithfield, Ltd., that was engaged in hog processing. The two groups charged the defendant with illegally discharging waste water from its plant into the Pagan River, near Smithfield, Virginia. The groups used the citizen suit provision under Section 505 of the Clean Water Act to allege violations of the statute, a law that also allows the court to levy civil penalties of up to $10,000 per day of violation.

In the first of a series of legal actions, CBF brought suit against Gwaltney of Smithfield, Ltd., in the district court. The judge in the case acknowledged that the CBF had standing to sue because its members in the area of the plant had been adversely affected. Despite the defendant's claims that the violations had occurred in the past, the judge agreed with the plaintiffs' argument that the violation was an ongoing one.

Judge Robert Merhige: In light of the language of the Clean Water Act's citizen suit provision, its legislative history, and its underlying policy goals, the Court concludes that the Clean Water Act authorizes citizen suits for civil penalties for violations of the Act, regardless of whether the polluter is engaged in unlawful conduct at the time the suit is filed or afterward. . . .

Source: *Chesapeake Bay Foundation v. Gwaltney of Smithfield, Ltd.* 611 F. Supp. 1542 (1984).

The judge then assessed a total of $1,285,322 in civil penalties against Gwaltney for problems stemming from its biological treatment system and violations related to the plant's chlorination system. The U.S. Supreme Court, however, disagreed with the district court and in 1987 ruled that the Clean Water Act only gave citizen groups jurisdiction against violations that were ongoing or likely to recur, as this section of the court's opinion indicates.

Justice Thurgood Marshall delivered the opinion of the Court: The most natural reading of Section 505 is a requirement that citizens-plaintiff allege a state of either continuous or intermittent violation—that is, a reasonable likelihood that a past polluter will continue to pollute in the future. . . . The interest of the citizen-plaintiff is primarily forward-looking. . . . Our conclusion that Section 505 does not permit citizen suits for wholly past violations does not necessarily dispose of this lawsuit. . . . The District Court found persuasive the fact that the citizens' allegation in the complaint that Gwaltney was continuing to violate its NPDES permit when plaintiffs filed suit, appears to have been made fully in good faith.

In the remand to the district court, the citizen-plaintiffs can prove an ongoing violation either 1) by proving actual violations that continue on or after the date the complaint is filed, or 2) by adducing evidence from which a reasonable trier of fact could find a continuing likelihood of a recurrence in intermittent or sporadic violations. Intermittent or sporadic violations do not cease to be ongoing until the date when there is no real likelihood of repetition."

Source: *Chesapeake Bay Foundation, Natural Resources Defense Council v. Gwaltney of Smithfield, Ltd.* 484 U.S. 49 (1987).

The next step in the litigation, after the U.S. Supreme Court's ruling, was to have the case reargued in 1988 in the same district court where the plaintiffs had originally filed their citizen suit. The judge once again ruled in favor of the two environmental groups, noting that there was a continuing likelihood of a recurrence of the plant's violations of the Clean Water Act and reinstating the original civil penalties. In 1989, the U.S. Court of Appeals supported the district court judge's ruling but lowered the amount of the civil penalties. The judges ruled that since some of the statutory violations had been corrected, the amount of the fines should be adjusted accordingly.

In 1997, a second Smithfield plant was found to have polluted the Pagan River with its discharges, but the state basically ignored the violations. The plant's owners threatened to move their operations to North Carolina, they cut costs by leaving its wastewater treatment plant unsupervised for one shift per day, and the plant's wastewater manager destroyed discharge records. CBF and NRDC put pressure on the Environmental Protection Agency to file its own suit against the company in the U.S. District Court. Here, the court ruled in favor of the EPA, assessing civil penalties for the violation of the Clean Water Act permit regulations.

Judge Rebecca Smith: When it became apparent that the Commonwealth's actions were not resulting in compliance, and the Commonwealth did not intend to seek a civil penalty for the violations, the EPA initiated its own enforcement action. . . .

During the entire trial, the defendants' approach to their Permit violations was rather cavalier. They repeatedly argued there was no real harm caused by their numerous violations, and that the Pagan River would still be environmentally damaged and unsafe for swimming and shellfish harvesting, even if defendants complied with their permit. Such arguments miss the mark. . . . A violator cannot escape liability or penalties for permit violations simply by pointing to the violations of others. Each must do their part to clean up the environment. . . .

Most of the defendants' permit discharge exceedances clearly had a severe and significant impact on the water quality of the Pagan River, in light of their frequency and severity. The harm to the environment and the risk to human health caused by defendants' numerous effluent limit violations are serious. . . . Accordingly, the court finds that the appropriate civil penalty for defendants' permit violations is $12,600,000.

Source: *US v. Smithfield Foods, et al.* 965 F. Supp. 769.

Mountain States Legal Foundation, et al. v. George W. Bush (2001)

During his second term as president, Bill Clinton used his power under the Antiquities Act of 1906 to issue proclamations that designated more than twenty wilderness and scenic areas as national monuments. The act does not require the president to consult with Congress or any other government agency, although there are limitations. The president may issue a proclamation to protect historic landmarks, historic and prehistoric structures, and other objects of historic and scientific interest. The Antiquities Act also requires that the area of land reserved by the president be the smallest area compatible with the proper care and management of the object that the president seeks to protect.

Although the proclamations issued by Clinton pleased many environmental groups who had sought protection for various sites, the president's actions angered others who believed he exceeded his authority as president. In this case, the Mountain States Legal Foundation, on behalf of itself and the off-road vehicle group the Blue Ribbon Coalition, filed suit, challenging Clinton's authority, as well as that of all future presidents, to establish and protect national monuments. They argued that such power belonged only to Congress. President George W. Bush is named as the defendant in this case, representing the federal government.

A coalition of environmental defense groups, represented by Earthjustice and the Wilderness Society, served as intervenors in the lawsuit and presented oral arguments to the U.S. District Court. They sought to have the lawsuit dismissed, arguing that the courts had a long history of affirming the president's power to establish monuments with significant natural values. Judge Paul Freedman agreed and dismissed the lawsuit as being without merit. This portion of the transcript of the case, which was heard 15 November 2001, outlines the arguments posed by each side, and the judge's ruling.

Michael Gheleta, for the defendant: Presidential authority exercised under the Antiquities Act has been subjected to judicial challenge some eight times, including recently, and in each of those instances the exercise of that authority was upheld.

The Supreme Court has addressed it on three occasions: in the Cameron case upholding the President's authority to designate Grand Canyon National Monument, the Cappaert case in 1976 involving Devil's Hole National Monument, and finally in United States versus California involving expansion of Channel Islands National Monument which involved islands out in the ocean. . . . Were plaintiffs to prevail in this case it would undermine nearly a century of Presidential exercise of authority under the Antiquities Act as well as jurisprudence upholding the exercise of that authority.

James S. Angell, for the intervenors: We've had 95 years of experience with the Antiquities Act. Presidents have protected Denali, Zion, Grand Teton, Glacier Bay, Grand Canyon, all to protect the natural, scientific and historic aspects of those areas. In fact, the very first monument created by President Roosevelt was the Devil's Tower in Wyoming, created because of its natural aspects.

Now Congress has addressed it several times in the years since this legislation but they've never ever questioned the President's power or limited it in any way to use the Antiquities Act to protect natural objects. And as the cases that we've cited point out, when you have a long, very visible history of action that is a statute, by an Executive, and Congress does nothing to rein it in or correct that interpretation, that's awfully strong evidence that that's the correct interpretation.

So I think this is a simple argument, Your Honor. Creating National Monuments to protect natural objects is within the statutory power and statutory language. It's been recognized by the Supreme Court. It's never been rejected by Congress. You shouldn't reject this claim to undo centuries of work.

Amanda Koehler, for the plaintiffs: The President maybe has more discretion because he is the President, but I just want to point out it doesn't change the fact that there are these limitations on his discretion in the Antiquities Act. Yes, the discretion was granted to him and he is the President, but the limitations are in the Act and the Government has admitted that there are limitations on his discretion under the Antiquities Act and we're either going to enforce these limitations or we're not, and that's the issue here.

Judge Paul L. Friedman, United States District Court Judge: By comparing the language of the statute of the Antiquities Act with the language of the Proclamation which is—and the Proclamations in this case are quite detailed—I find that the President met his responsibilities to the Congress under the Act and that it's not my role to look at the specific decisions so long as he applied those standards.

Source: U.S. District Court for the District of Columbia, Docket No. CA 00–2072, 15 November 2001.

7

Organizations

Among the thousands of organizations with an interest in the environment are those with a local issue and constituency, species- or issue-specific groups, regional and state groups that are often branches or chapters of a larger interest group, highly professionalized law and research centers, trade associations, well-established national and global organizations, nonprofit groups, and others that have traded their charitable status in order to actively affect the political system.

This chapter provides a sampling of each type of group, all of which have an environmental agenda and a base of activist members. They represent a wide spectrum of organizations across the United States, along with several that work on a global level. As with any organization, it is important to remember that some groups are transient and information about them is subject to change.

Abalone Alliance Safe Energy Clearinghouse
2940 16th Street, #310
San Francisco, CA 94103
Telephone: (415) 861-0592
Fax: (415) 558-8135
Web site: http://www.energy-net.org

The Abalone Alliance works toward the use of renewable energy sources in northern California while simultaneously opposing the construction of nuclear reactors.

African Wildlife Foundation
1400 16th Street NW, Suite 120
Washington, DC 20036
Telephone: (202) 939-3333
Fax: (202) 939-3332
Web site: http://www.awf.org

With an office in Nairobi, Kenya, this U.S.-based organization works to conserve the wild animals of Africa and their habitat.

'Ahahui Mālama I Ka Lōkahi
P.O. Box 61578
Honolulu, HI 96839-1578
Web site: http://www.aecos.com/aml

Created by native Hawaiians who recognize that the state's unique native plants, animals, and ecosystems represent a vital cultural resource in danger of extinction, the group conducts research and issues position papers on resource preservation.

Air and Waste Management Association
1 Gateway Center, Third Floor
Pittsburgh, PA 15222
Telephone: (412) 232-3444
Fax: (412) 232-3450
Web site: http://www.awma.org

This trade association brings together professionals and researchers with an interest in air quality and waste issues and provides updates on new legislation and regulations.

Akwesasne Task Force on the Environment
P.O. Box 992
Hogansburg, NY 13655
Telephone: (518) 358-9607
Fax: (518) 358-2857
Web site: http://www.slic.com/atfe/atfe.htm

The Akwesasne Task Force on the Environment is a community-based group organized to address the environmental problems facing the Mohawk Nation.

Alaska Conservation Foundation
441 West 5th Avenue, Suite 402

Anchorage, AK 99501
Telephone: (907) 274-1917
Fax: (907) 274-4145
Web site: http://www.akcf.org

Alaska Conservation Foundation supports grassroots activism by emphasizing collaboration among conservation organizations and public education programs working on local and national issues.

Alliance for Sustainable Jobs and the Environment
1125 SE Madison
Portland, OR 97214
Telephone: (503) 736-9777
Fax: (503) 736-9776
Web site: http://www.asje.org

In an unusual partnership between environmental activists and labor leaders, the group seeks to forge trade policies that protect jobs and the environment, identifying rogue companies that do not do so as well as model companies that do.

Alliance for the Wild Rockies
P.O. Box 8731
Missoula, MT 59807
Telephone: (406) 721-5420
Fax: (406) 721-9917
Web site: http://www.wildrockiesalliance.org

Alliance for the Wild Rockies works to sustain the ecological integrity of the Rockies through citizen education, conservation biology, and environmental law.

Alliance to Save Energy
1200 18th Street NW, Suite 900
Washington, DC 20036
Telephone: (202) 857-0666
Fax: (202) 331-9588
Web site: http://www.ase.org

The Alliance to Save Energy works to promote efficient and clean energy sources to benefit the environment, society, and economic and national security interests.

Amazon Watch
115 South Topanga Canyon Boulevard, Suite E
Topanga, CA 90290
Telephone: (310) 455-0617
Fax: (310) 455-0619
Web site: http://amazonwatch.org

Working with indigenous and environmental organizations in the
Amazon Basin, this group provides information and advocacy for
the region and its native people.

American Chemical Society
1155 16th Street NW
Washington, DC 20036
Telephone: (202) 872-4600
Fax: (202) 872-4615
Web site: http://www.acs.org

One of the oldest trade associations in the country, the group has
had a profound influence on a broad spectrum of environmental
issues.

American Council for an Energy-Efficient Economy
1001 Connecticut Avenue NW, Suite 801
Washington, DC 20036
Telephone: (202) 429-8873
Fax: (202) 429-2248
Web site: http://www.aceee.org

The American Council for an Energy-Efficient Economy works to
advance energy efficiency for the benefit of the economy and the
environment as a means of promoting both economic prosperity
and environmental protection.

American Farmland Trust
1200 18th Street NW, Suite 800
Washington, DC 20036
Telephone: (202) 331-7300
Fax: (202) 659-8339
Web site: http://www.farmland.org

American Farmland Trust works to decrease the loss of farmland
and helps promote farming practices that are healthy for citizens
and the environment.

American Fisheries Society (AFS)
5410 Grosvenor Lane, Suite 110
Bethesda, MD 20814
Telephone: (301) 897-8616
Fax: (301) 897-8096
Web site: http://www.fisheries.org

AFS promotes scientific research to help in the conservation and sustainable development of fisheries. AFS represents the largest number of fisheries scientists in the United States, encourages continuing education for those scientists, and publishes leading fisheries journals to educate its members and the public.

American Land Rights Association (ALRA)
P.O. Box 400
Battle Ground, WA 98604
Telephone: (360) 687-2471
Fax: (360) 687-2973
Web site: http://www.landrights.org

An umbrella organization for several projects, the ALRA includes the League of Private Property Voters (LPPV). Using a format similar to that of the League of Conservation Voters, the LPPV judges how each member of Congress voted on a number of private property, multiple use, or resource development issues.

American Oceans Campaign
600 Pennsylvania Avenue SE, Suite 210
Washington, DC 20003
Telephone: (202) 544-3526
Fax: (202) 544-5625
Web site: http://www.americanoceans.org

This group alerts its members and the public to issues related to the protection and vitality of America's water and ocean fish.

American Petroleum Institute
1220 L Street NW
Washington, DC 20005
Telephone: (202) 682-8000
Fax: (202) 682-8154
Web site: http://www.api.org

A major trade organization representing the oil and natural gas industries of the United States, including exploration, transportation, refining, and marketing.

American Recreation Coalition
1225 New York Avenue NW, Suite 450
Washington, DC 20005
Telephone: (202) 682-9530
Fax: (202) 682-9529
Web site: http://www.funoutdoors.com

Sponsored by a large number of recreational users, including recreation vehicle dealers, the American Horse Council, and the Motorcycle Industry Council, the organization is dedicated to the protection and enhancement of everyone's right to health and happiness through recreation.

American Rivers
1025 Vermont Avenue NW, Suite 720
Washington, DC 20005
Telephone: (202) 347-7500
Fax: (202) 347-9240
Web site: http://www.amrivers.org

American Rivers is devoted to protecting and restoring rivers nationwide. It also seeks to foster a river stewardship ethic and addresses issues ranging from flood control and hydropower policy reform to urban rivers and riparian species protection.

American Society for the Prevention of Cruelty to Animals
424 East 92nd Street
New York, NY 10128
Telephone: (212) 876-7700
Fax: (212) 410-7658
Web site: http://www.aspca.org

The American Society for the Prevention of Cruelty to Animals works to educate the public about animal welfare issues, including animal cruelty and suffering, through public outreach and legislative programs.

American Solar Energy Society
2400 Central Avenue, Suite G-1

Boulder, CO 80301
Telephone: (303) 443-3130
Fax: (303) 443-3212
Web site: http://www.ases.org

The American Solar Energy Society is dedicated to increasing the use of solar energy for the benefit of energy consumers and the global environment. The group has chapters throughout the United States.

American Wind Energy Association
122 C Street NW, Suite 380
Washington, DC 20001
Telephone: (202) 383-2500
Fax: (202) 383-2505
Web site: http://www.awea.org

The American Wind Energy Association advocates the development of wind energy as a reliable and environmentally friendly energy alternative in both the United States and the world.

Amigos Bravos
P.O. Box 238
Taos, NM 87571
Telephone: (505) 758-3874
Fax: (505) 758-7345
Web site: http://www.amigosbravos.org

State and local activists seek to force the cleanup of waste from the Molycorp mine near the confluence of the Rio Grande and the Red River in a debate that has pitted residents who depend upon the mine's jobs against environmental organizations.

Antarctica Project
1630 Connecticut Avenue NW, Third Floor
Washington, DC 20009
Telephone: (202) 234-2480
Fax: (202) 387-4823
Web site: http://www.asoc.org

This is the only conservation organization in the world that works exclusively in leading the domestic and international campaigns to protect Antarctica's pristine wilderness and environment.

Atlantic Salmon Federation
P.O. Box 807
Calais, ME 04619-0807 or

P.O. Box 5200
St. Andrews, New Brunswick
Canada E5B 3S8
Telephone: (506) 529-1033
Fax: (506) 529-4438
Web site: http://www.asf.ca

With offices in Canada and the United States, the Atlantic Salmon Federation is an example of an organization that deals with the regional issues of overfishing, fish farms, and fishing exclusion zones.

Beyond Pesticides
701 E Street SE, #200
Washington, DC 20003
Telephone: (202) 543-5450
Fax: (202) 543-4791
Web site: http://www.beyondpesticides.org

Beyond Pesticides is a group dedicated to educating citizens about pesticides and nontoxic alternatives.

Blue Ribbon Coalition
4555 Burley Drive, Suite A
Pocatello, ID 83202
Telephone: (208) 237-1008
Fax: (208) 237-9424
Web site: http://sharetrails.org

The Blue Ribbon Coalition champions multiple use of public lands for the benefit of motorized and other recreationists by educating and empowering its members.

Bluewater Network
300 Broadway, Suite 28
San Francisco, CA 94133
Telephone: (415) 788-3666
Fax: (415) 788-7324
Web site: http://www.bluewaternetwork.org

The Bluewater Network is a project of Earth Island Institute working specifically on the causes of climate change and fighting damage caused by the oil transport and motorized recreation industries.

Boreal Forest Network
2–70 Albert Street
Winnipeg, Manitoba
Canada R3B 1E7
Telephone: (204) 947-3081
Fax: (204) 947-3076
Web site: http://www.borealnet.org

Founded in 1994, the Boreal Forest Network's mission is the protection, restoration, and sustainable use of the boreal forest worldwide.

Brown Bear Resources (BBR)
222 North Higgins
Missoula, MT 59802
Telephone: (406) 549-4896
Fax: (406) 549-4884
Web site: http://www.brownbear.org

BBR focuses its efforts in and around the Northern Continental Divide Ecosystem of northwest Montana. Its activities involve educating humans about their role in determining the survival of the grizzly bear, as well as many other species that also depend upon the grizzlies' habitat for survival.

California Oak Foundation (COF)
1212 Broadway, Suite 810
Oakland, CA 94612
Telephone: (510) 763-0282
Fax: (510) 208-4435
Web site: http://www.californiaoaks.org

Founded in 1988, the COF is committed to preserving California's oak forest ecosystem and its rural landscapes by advocating curbs on urban sprawl and creating livable cities.

Campaign for Safe and Affordable Drinking Water
4455 Connecticut Avenue NW, Suite A-300
Washington, DC 20008-2328

Telephone: (202) 895-0420
Fax: (202) 895-0438
Web site: http://www.safe-drinking-water.org

This alliance of environmental and public health organizations is working to educate the public about drinking water safety and standards, improvement, and protection.

Cascadia Forest Alliance
P.O. Box 4946
Portland, OR 97208
Telephone: (503) 241-4879
Fax: Same as telephone number
Web site: http://www.cascadiaforestalliance.org

Cascadia Forest Alliance works with nonviolent direct action to protect regional forests.

Center for Biological Diversity (CBD)
P.O. Box 710
Tucson, AZ 85702
Telephone: (520) 623-5252
Fax: (520) 623-9797
Web site: http://www.biologicaldiversity.org

The CBD dedicates its work to protecting endangered species and wild places through policy, environmental law, science, and education.

Center for Community Action and Environmental Justice
P.O. Box 33124
Riverside, CA 92519
Telephone: (909) 360-8451
Fax: (909) 360-5950
Web site: http://www.ccaej.org

The center's goal is to bring groups of people together to find opportunities for cooperation, agreement, and problem solving by developing democratically based, participatory organizations that promote involvement of a diverse segment of the community.

Center for Environmental Citizenship
200 G Street NE, Suite 300
Washington, DC 20002

Telephone: (202) 547-8435
Fax: (202) 547-8572
Web site: http://www.envirocitizen.org

This group is an umbrella organization for a number of programs, including Campus Green Vote, designed to network young leaders to protect the environment.

Center for Food Safety (CFS)
660 Pennsylvania Avenue SE, Suite 302
Washington, DC 20003
Telephone: (202) 547-9359
Fax: (202) 547-9429
Web site: http://www.centerforfoodsafety.org

The CFS provides leadership in legal, scientific, and grassroots efforts to address the increasing concerns about the impacts of our food production system on human health, animal welfare, and the environment.

Center for Health, Environment, and Justice
P.O. Box 6806
Falls Church, VA 22040
Telephone: (703) 237-2249
Fax: (703) 237-8389
Web site: http://www.chej.org

The Center for Health, Environment, and Justice works to promote a clean and healthy environment for all people, regardless of their race or economic standing.

Center for Indigenous Environmental Resources
245 McDermot Avenue, Third Floor
Winnipeg, Manitoba
Canada R3B 0S6
Telephone: (204) 956-0660
Fax: (204) 956-1895
Web site: http://www.cier.mb.ca

The Center for Indigenous Environmental Resources provides educational information regarding present-day environmental issues that threaten indigenous cultures.

Center for Marine Conservation
1725 DeSales Street NW, Suite 600
Washington, DC 20036
Telephone: (202) 429-5609
Fax: (202) 872-0619
Web site: http://www.cmc-ocean.org

This membership organization is dedicated to conserving coastal and ocean resources and to protecting marine wildlife and habitats through science-based advocacy, research, and public education.

Center for the Defense of Free Enterprise
12500 NE Tenth Place
Bellevue, WA 98005
Telephone: (425) 455-5038
Fax: (425) 451-3959
Web site: http://www.cdfe.org

Founded in 1976, the center is a nonpartisan education and research organization that works on free enterprise studies, public policy research, book publishing, conferences, white papers, and media outreach.

Circumpolar Conservation Union (CCU)
600 Pennsylvania Avenue SE, Suite 210
Washington, DC 20003
Telephone: (202) 675-8370
Fax: (202) 675-8373
Web site: http://www.circumpolar.org

CCU is dedicated to protecting the ecological and cultural integrity of the Arctic and promoting cooperation among Arctic peoples, environmental organizations, and other interests. Its goal is to achieve a comprehensive legal and policy regime for the Arctic.

Citizens for a Better Environment
152 West Wisconsin Avenue, #510
Milwaukee, WI 53203
Telephone: (414) 271-7280
Fax: (414) 271-5904
Web site: http://www.cbemw.org

Citizens for a Better Environment works primarily in the Great Lakes region developing strategies for improving environmental quality of the area.

Clean Water Action
4455 Connecticut Avenue NW, Suite A300
Washington, DC 20008-2328
Telephone: (202) 895-0420
Fax: (202) 895-0438
Web site: http://www.cleanwateraction.org

Clean Water Action advocates the availability of clean, affordable, and healthy drinking water for all citizens.

Clean Water Network
1200 New York Avenue NW, Suite 400
Washington, DC 20005
Telephone: (202) 289-2395
Fax: (202) 289-1060
Web site: http://www.cwn.org

This is an alliance of more than 1,000 organizations that support strong clean water safeguards to protect human health and the environment.

Coalition for Environmentally Responsible Economies
11 Arlington Street, Sixth Floor
Boston, MA 02116
Telephone: (617) 247-0700
Fax: (617) 267-5400
Web site: http://www.ceres.org

The Coalition for Environmentally Responsible Economies works to standardize corporate environmental reporting and promotes the transformation of environmental management within firms.

Colorado Wild
P.O. Box 2434
Durango, CO 81302
Telephone: (970) 385-9833
Fax: (970) 259-8303
Web site: http://www.coloradowild.org

Arguing that public lands are being exploited in ski areas, the group has opposed government approval that allows snow making using wastewater under the provisions of the Clean Air Act.

Communities for a Better Environment
1611 Telegraph Avenue, Suite 450
Oakland, CA 94612
Telephone: (510) 302-0430
Fax: (510) 302-0437
Web site: http://www.cbecal.org

This environmental health and justice organization promotes clean air, clean water, and the development of toxin-free communities through grassroots activism, environmental research, and legal assistance within underserved urban communities.

Conservation Fund
1800 North Kent Street, Suite 1120
Arlington, VA 22209
Telephone: (703) 525-6300
Fax: (703) 525-4610
Web site: http://www.conservationfund.org

The Conservation Fund aids local, state, and federal agencies in acquiring and preserving open spaces.

Conservation International
1919 M Street NW, Suite 600
Washington, DC 20036
Telephone: (202) 912-1000
Fax: (202) 912-1030
Web site: http://www.conservation.org

Through environmental education and advocacy, Conservation International promotes the conservation of global biodiversity and the notion that human societies can live harmoniously with nature.

Co-op America
1612 K Street NW, Suite 600
Washington, DC 20006
Telephone: (202) 872-5307
Fax: (202) 331-8166
Web site: http://www.coopamerica.org

Co-op America provides economic and organizing ideas so that businesses and individuals can effectively address today's social and environmental problems.

CorpWatch
P.O. Box 29344
San Francisco, CA 94129
Telephone: (415) 561-6568
Fax: (415) 561-6493
Web site: http://www.corpwatch.org

CorpWatch encourages grassroots education and activism to counter international globalization efforts.

Council for Responsible Genetics
5 Upland Road, Suite 3
Cambridge, MA 02140
Telephone: (617) 868-0870
Fax: (617) 491-5344
Web site: http://www.gene-watch.org

Devoted to fostering public debate about the social, ethical, and environmental implications of new genetic technologies, the organization monitors new developments and their applications to human genetics and commercial biotechnology.

Defenders of Wildlife
1101 14th Street NW, #1400
Washington, DC 20005
Telephone: (202) 682-9400
Fax: (202) 682-1331
Web: http://www.defenders.org

Defenders of Wildlife is dedicated to protecting all native wild animals and plants in their natural communities.

Dogwood Alliance
P.O. Box 7645
Asheville, NC 28802
Telephone: (828) 251-2525
Fax: (828) 251-2501
Web site: http://www.dogwoodalliance.org

Dogwood Alliance is a network of grassroots environmental groups working to protect southern forests and communities.

Downwinders
P.O. Box 111
Lava Hot Springs, ID 83246-0111
Telephone: None
Web site: http://www.downwinders.org

This is a research and educational foundation that seeks to expose the plight of downwind residents whose exposures to nuclear fallout have caused illnesses, to obtain justice for their injuries, and to seek an immediate end to nuclear testing.

Earth Day Network
91 Marion Street
Seattle, WA 98104
Telephone: (206) 876-2000
Fax: (206) 682-1184
Web site: http://www.earthday.net

This national organization organizes and coordinates Earth Day observances and seeks to increase public awareness about environmental issues.

Earth Island Institute
300 Broadway, Suite 28
San Francisco, CA 94133
Telephone: (415) 788-3666
Fax: (415) 788-7324
Web site: http://www.earthisland.org

Founded by environmentalist David Brower, the Earth Island Institute supports projects that promote the conservation, preservation, and restoration of the earth.

Earthjustice Legal Defense Fund
180 Montgomery Street, Suite 1400
San Francisco, CA 94104-4209
Telephone: (415) 627-6700
Fax: (415) 627-6740
Web site: http://www.earthjustice.org

Earthjustice practices public interest law dedicated to protecting natural resources and wildlife as well as promoting the right of all people to a healthy environment.

Earth Share
3400 International Drive NW, Suite 2K
Washington, DC 20008
Telephone: (800) 875-3863
Fax: (202) 537-7101
Web site: http://www.earthshare.org

This group enables individuals to support environmental organizations by managing workplace giving campaigns for its member charities.

EarthSave International
1509 Seabright Avenue, Suite B1
Santa Cruz, CA 95062
Telephone: (831) 423-0293
Fax: (831) 423-1313
Web site: http://www.earthsave.org

EarthSave promotes food choices that are healthy for people and the planet.

EarthWatch
3 Clock Tower Place, Suite 100
Box 75
Maynard, MA 01754
Telephone: (978) 461-0081
Fax: (978) 461-2332
Web site: http://www.earthwatch.org

EarthWatch Institute promotes sustainable conservation of natural resources and cultural heritage by facilitating partnerships among scientists, educators, and the general public.

Eastern Native Seed Conservancy
P.O. Box 451
Great Barrington, MA 01230
Telephone: (413) 229-8316
Fax: None
Web site: http://www.berkshire.net/ensc/seedmain.html

Eastern Native Seed Conservancy works to preserve rare seeds in order to save these cultural resources from extinction.

Ecological Society of America
1707 H Street NW, Suite 400
Washington, DC 20006
Telephone: (202) 833-8773
Fax: (202) 833-8775
Web site: http://www.esa.org

These scientists conduct research, teach, and work to provide the ecological knowledge needed to solve environmental problems.

Environmental and Energy Study Institute
122 C Street NW, Suite 700
Washington, DC 20001
Telephone: (202) 628-1400
Fax: (202) 628-1825
Web site: http://www.eesi.org

The Environmental and Energy Study Institute works on the development of public policies that are healthy for people and the environment.

Environmental Defense
257 Park Avenue South
New York, NY 10010
Telephone: (212) 505-2100
Fax: (212) 505-2375
Web site: http://www.environmentaldefense.org/home.cfm

Committed to both domestic and international issues, Environmental Defense brings together social and scientific experts to address wide-reaching environmental issues that affect oceans, natural resources, and wildlife. The group maintains an on-line Action Alert list of 750,000 members with information about pending government actions.

Environmental Law Institute (ELI)
1616 P Street NW, Suite 200
Washington, DC 20036
Telephone: (202) 939-3800

Fax: (202) 939-3868
Web site: http://www.eli.org

The ELI comprises international and domestic lawyers and environmental practitioners from government, the private bar, citizen organizations, and businesses seeking to keep informed on current issues and disputes.

Environmental Support Center
1500 Massachusetts Avenue NW, Suite 25
Washington, DC 20005
Telephone: (202) 331-9700
Fax: (202) 331-8592
Web site: http://www.envsc.org

Environmental Support Center empowers the many grassroots organizations in the United States by helping to improve their management, planning, and outreach capabilities.

Environmental Working Group (EWG)
1718 Connecticut Avenue NW, Suite 600
Washington, DC 20009
Telephone: (202) 667-6982
Fax: (202) 232-2592
Web site: http://www.ewg.org

Through research and information dissemination, EWG provides the public with locally relevant information on environmental issues ranging from pesticide use and drinking water safety to air pollution and toxic substances.

Evangelical Environmental Network USA
10 East Lancaster Avenue
Wynnewood, PA 19096-3495
Telephone: (610) 645-9390
Web site: http://cesc.montreat.edu/ceo/EEN/EENusa.html

This organization was initiated by World Vision and Evangelicals for Social Action as part of a growing movement to develop a biblical response to the disregard of God's creation. It was formed to respond to the recognition of many of the world's scientists that environmental problems are at their roots spiritual problems and require a response grounded in faith.

Forest Ecology Network (FEN)
P.O. Box 2118
Augusta, ME 04338
Telephone: (207) 628-6404
Fax: (207) 628-5741
Web site: http://www.powerlink.net

The purpose of the FEN is to protect the native forest environment of Maine through public awareness, grassroots citizen activism, and education.

Forest Guardians
312 Montezuma, Suite A
Santa Fe, NM 87501
Telephone: (505) 988-9126
Fax: (505) 989-8623
Web site: http://www.fguardians.org

Sometimes considered one of the most philosophically radical environmental groups, Forest Guardians opposes all attempts to log forests.

Forest Service Employees for Environmental Ethics
P.O. Box 11615
Eugene, OR 97440
Telephone: (541) 484-2692
Fax: (541) 484-3004
Web site: http://www.fseee.org

This group, whose members are U.S. Forest Service employees and citizens, works to protect the remaining forest wilderness from logging and overuse.

Forest Trust
P.O. Box 519
Santa Fe, NM 87504
Telephone: (505) 983-8992
Fax: (505) 986-0798
Web site: http://theforestrust.org

The Forest Trust works to achieve a middle ground in the forest management debate. The trust works collaboratively with interested stakeholders to forge new and creative environmental solutions.

Friends of the Earth (FOE)
1025 Vermont Avenue NW
Washington, DC 20005
Telephone: (202) 783-7400
Fax: (202) 783-0444
Web site: http://www.foe.org

Friends of the Earth's work to help protect the earth's finite resources is in three main program areas: economic issues, international advocacy, and community organizing.

Friends of the Earth International
P.O. Box 19199
1000gd Amsterdam
The Netherlands
Telephone: +31 20 622 1369
Fax: +31 20 639 2181
Web site: http://www.foei.org

The international secretariat of FOE coordinates campaigns throughout the world, including preservation of natural resources, participation by indigenous peoples, and trade and economic assistance.

Friends of the Everglades
7800 Red Road, Suite 215K
Miami, FL 33143
Telephone: (305) 669-0858
Fax: (305) 669-4108
Web site: http://www.everglades.org

Friends of the Everglades works to protect, restore, and preserve the Greater Kissimee–Okeechobee–Everglades ecosystem.

Friends of the River
915 20th Street
Sacramento, CA 95814
Telephone: (916) 442-3155
Fax: (916) 442-3396
Web site: http://www.friendsoftheriver.org

Using technical expertise and campaign work restoring rivers, this group focuses on federal hydropower licensing.

Fund for Animals
200 West 57th Street
New York, NY 10019
Telephone: (212) 246-2096
Fax: (212) 246-2633
Web site: http://www.fund.org

Founded in 1967 by activist Cleveland Amory, the Fund for Animals' motto is "We speak for those who can't." The group opposes sport hunting, commercial trapping, and acts of cruelty toward animals as one of the nation's major animal protection organizations.

Geological Society of America (GSA)
3300 Penrose Place
Boulder, CO 80301
Telephone: (303) 447-2020
Fax: (303) 357-1070
Web site: http://www.geosociety.org

GSA is an information storehouse for earth scientists in all sectors of society: academic, government, business, and industry.

Get Oil Out! (GOO!)
914 Anacapa Street
Santa Barbara, CA 93102
Telephone: (805) 961-3968
Web site: http://www.getoilout.org

Formed as a result of the 1969 oil spill off the coast of Santa Barbara, California, GOO! is one of the nation's oldest grassroots organizations, formed to protect the coastline.

Global Response
P.O. Box 7490
Boulder, CO 80306-7490
Telephone: (303) 444-0306
Fax: (303) 449-9794
Web site: http://www.globalresponse.org

Global Response is an environmental action and education network that encourages citizens to contact individuals and companies involved in environmentally destructive projects around the world.

Goldman Environmental Foundation
One Lombard, Suite 303
San Francisco, CA 94111
Telephone: (415) 788-1090
Fax: (415) 788-7890
Web site: http://www.goldmanfund.org

For over fifty years, the Goldman family has provided seed money and grants to individuals, such as their contribution of $750,000 to the World Wildlife Fund's Global Toxic Chemicals Initiative.

Grand Canyon Trust (GCT)
2601 North Fort Valley Road
Flagstaff, AZ 86001
Telephone: (928) 774-7488
Fax: (928) 774-7570
Web site: http://www.grandcanyontrust.org

GCT protects and restores the canyon country of the Colorado Plateau, including projects such as restoring the Virgin River and monitoring Clean Air Act violations at the San Juan nuclear power facility.

Great Basin Mine Watch
P.O. Box 10262
Reno, NV 89510
Telephone: (775) 348-1986
Fax: (775) 324-2677
Web site: http://www.greatbasinminewatch.org

Activists from this group monitor mining within the region, such as a clay mining operation proposed by the world's largest kitty-litter producer near Reno, Nevada.

Greenaction for Health and Environmental Justice
One Hallidie Plaza, Suite 760
San Francisco, CA 94102
Telephone: (415) 248-5010
Fax: (415) 248-5011
Web site: http://www.greenaction.org

Greenaction works with community members to effect change and alter government and corporate policies.

GreenMarketplace.com
5801 Beacon Street, Suite 2
Pittsburgh, PA 15217
Telephone: (412) 420-6400 or (888) 59-EARTH
Fax: (412) 420-6404
Web site: http://www.greenmarketplace.com

GreenMarketplace offers consumers environmentally and socially responsible alternatives for household products, clothing, home and garden items, and food via the Internet.

Greenpeace International
Keizersgracht 176
1016 DW Amsterdam
The Netherlands
Telephone: +31 20 523 6222
Fax: +31 20 523 6200
Web site: http://www.greenpeace.org

With branches in both industrialized and developing countries, Greenpeace International serves as the hub of an international organization with a broad environmental agenda focused on marine conservation, pesticides, and nuclear energy.

Greenpeace USA
702 H Street NW
Washington, DC 20001
Telephone: (800) 326-0959
Fax: (202) 462-4507
Web site: http://www.greenpeaceusa.org

Greenpeace uses nonviolent direct action and alternative communication to expose environmental problems and to promote solutions.

Gwich'in Steering Committee
122 1st Avenue, Box 2
Fairbanks, AK 99701
Telephone: (907) 458-8264
Fax: (907) 457-8265
Web site: http://www.alaska.net/~gwichin

Composed of tribal members from Alaska and Canada, the steering committee represents the interests of the Gwich'in Nation, especially the protection of the Arctic National Wildlife Refuge and indigenous peoples' cultural survival.

Headwaters Environmental Center
84 Fourth Street
P.O. Box 729
Ashland, OR 97520
Telephone: (541) 482-4459
Fax: (541) 482-7282
Web site: http://www.headwaters.org

Headwaters works to protect watersheds, wildlife, and biological diversity in southwest Oregon.

Heart of America Northwest
1305 Fourth Street, Cobb Center, Suite 208
Seattle, WA 98101
Telephone: (206) 382-1014
Fax: (206) 382-1148
Web site: http://www.heartofamericanorthwest.org

Heart of America's mission is to advance the health of the environment in the Northwest. The group is one of the leading grassroots organizations working on the cleanup of the Hanford Nuclear Reservation.

Heartwood
P.O. Box 1424
Bloomington, IN 47402
Telephone: (812) 337-8898
Fax: (812) 337-8892
Web site: http://www.heartwood.org

Citizen forest activists from four states (Indiana, Illinois, Ohio, and Kentucky) joined together in 1990–1991 to use a regional approach to projects to restore the temperate hardwood forest and to show support for ending logging in the area's public forests.

High North Alliance
P.O. Box 123
N-8398 Reine i Lofoten
Norway
Telephone: 47 76 09 24 14

Fax: 47 76 09 24 50
Web site: http://www.highnorth.no/Default.asp

The objective of this group is to defend the right of coastal communities to utilize marine mammals sustainably, especially whales.

Indigenous Environmental Network
P.O. Box 485
Bemidji, MN 56619
Telephone: (218) 751-4967
Fax: (218) 751-0561
Web site: http://www.ienearth.org

The Indigenous Environmental Network is an alliance of grassroots indigenous peoples organized to protect the earth from contamination and exploitation through respecting traditional teachings and natural laws.

International Rivers Network
1847 Berkeley Way
Berkeley, CA 94703
Telephone: (510) 848-1155
Fax: (510) 848-1008
Web site: http://www.irn.org

International Rivers Network networks with communities in the United States and worldwide to prevent the destruction of watersheds and free-flowing rivers, encouraging equitable and sustainable methods of meeting the needs for water, energy, and flood management.

Izaak Walton League of America
707 Conservation Lane
Gaithersburg, MD 20878
Telephone: (301) 548-0150
Fax: (301) 548-0146
Web site: http://www/iwla.org

The Izaak Walton League of America works to protect wildlands in the United States through advocating policy changes in legislation.

Jane Goodall Institute
P.O. Box 14890
Silver Spring, MD 20911

Telephone: (301) 565-0086
Fax: (301) 565-3188
Web site: http://www.janegoodall.org

The institute, founded by the renowned primate researcher, advances the power of individuals to take informed and compassionate action to improve the environment for all living things.

Kahea
P.O. Box 27112
Honolulu, HI 96827-0112
Web site: http://www.kahea.org

A network of activists throughout the five main Hawaiian islands, Kahea (which translates as "the call") serves as an alliance that convenes citizen participation and organizes focus campaigns to protect Hawaii's fragile environment.

Lands Council
517 South Division Street
Spokane, WA 99202
Telephone: (509) 838-4912
Fax: (509) 838-5155
Web site: http://www.landscouncil.org

The Lands Council works to protect the ecosystem of the inland Columbia River watershed by supporting healthy forests and watersheds.

League of Conservation Voters
1920 L Street NW, Suite 800
Washington, DC 20036
Telephone: (202) 785-8683
Fax: (202) 835-0491
Web site: http://www.lcv.org

The League of Conservation Voters works to protect the environment through political action, tracking the voting records of members of Congress on environmental issues and exposing politicians who vote against a healthy environment.

Legal Environmental Assistance Foundation, Inc.
1114 Thomasville Road, Suite E

Tallahassee, FL 32303
Telephone: (850) 681-2591
Fax: (850) 224-1275
Web site: http://www.leaf-envirolaw.org

The Legal Environmental Assistance Foundation is a public interest law firm working on issues of clean groundwater and surface water in Alabama, Florida, and Georgia.

Living Lands and Waters
17615 Great River Road
East Moline, IL 61244
Telephone: (309) 496-9848
Fax: (309) 496-1012
Web site: http://www.cleanrivers.com

This group aids in the protection, preservation, and restoration of the nation's major rivers and their watersheds, using volunteers to clean up river areas and to expand awareness of environmental issues and stewardship.

Living Rivers
P.O. Box 1589
Scottsdale, AZ 85252
Telephone: (480) 990-7839
Fax: (480) 990-2662
Web site: http://www.drainit.org

Living Rivers promotes large-scale river restoration by undoing the damage done by dams, diversions, and unmitigated pollution. The group engages in investigation, litigation, and demonstrations.

Maine Conservation Rights Institute
Bob Voight
Lubec, ME 04652
Telephone: (207) 733-5593
Fax: (207) 733-2014
Web site: http://www.mecri.org

Committed to the application of innovative conservation methods within the context of private property and free market systems, this organization seeks to inform the public on the rights of individual landowners.

Military Toxics Project (MTP)
P.O. Box 558
Lewiston, ME 04243
Telephone: (207) 783-5091
Fax: (207) 783-5096
Web site: http://www.miltoxproj.org

The mission of the MTP is to unite activists, organizations, and communities in the struggle to clean up military pollution, to safeguard the transportation of hazardous materials, and to advance the development and implementation of preventative solutions.

Moab Citizens Alliance
P.O. Box 1003
Moab, UT 84532
Telephone: (435) 259-1517
Fax: Same as telephone number
Web site: http://www.moabutah.org

Faced with a two-tier society of residents and tourists, the organization monitors new development and what it terms the "Aspenization" of the region.

Mothers' Environmental Coalition of Alabama
700 Eighth Avenue West
Birmingham, AL 35204
Telephone: (205) 254-8006
Fax: (205) 251-1877
Web site: http://www.bham.net/SOE/clepp

To protect children's health, the group focuses on childhood lead poisoning, air quality, pesticides, and other environmental issues.

Mountain States Legal Foundation (MSLF)
707 17th Street, Suite 3030
Denver, CO 80202
Telephone: (303) 292-2021
Fax: (303) 292-1980
Web site: http://www.mountainstateslegal.org

As a politically conservative nonprofit group, the MSLF is dedicated to individual liberty, the right to own and use property, limited government, and the free enterprise system.

National Audubon Society
700 Broadway
New York, NY 10003
Telephone: (212) 979-3000
Fax: (212) 979-3188
Web site: http://www.audubon.org

The National Audubon Society works on protecting natural ecosystems for the benefit of birds and other wildlife.

National Brownfield Association
3105-C North Wilke Road
Arlington Heights, IL 60004
Telephone: (847) 870-8208
Fax: (847) 870-8331
Web site: http://www.brownfieldsnet.org

This organization is a diverse group of interests that provides a unified voice for brownfields cleanup (a relatively new policy to clean up abandoned or vacant buildings, lots, and manufacturing sites for reuse for parks, housing, or commercial buildings, and to eliminate blight in urban areas) through a quarterly newsletter, annual meeting, legislative support, projects, and financing.

National Coalition for Marine Conservation
3 North King Street
Leesburg, VA 20176
Telephone: (703) 777-0037
Fax: (703) 777-1107
Web site: http://www.savethefish.org

This is the nation's oldest public advocacy group dedicated exclusively to conserving the world's ocean fish, habitat, and environment through public awareness of the threats to marine fisheries.

National Environmental Coalition of Native Americans
2213 West 8th Street
Prague, OK 74864
Telephone: (405) 567-4297
Fax: (405) 567-4297
Web site: http://www.alphacdc.com/necona

The National Environmental Coalition of Native Americans formed to educate Indians and non-Indians about the risks of radioactivity and to coordinate tribes to declare their nations nuclear-free zones.

National Environmental Health Association
720 South Colorado Boulevard, Suite 970
Denver, CO 80222
Telephone: (303) 756-9090
Fax: (303) 691-9490
Web site: http://www.neha.org

This group started as a professional society for environmental health practitioners. Today it offers various national credential programs and technical workshops and publishes its own journal for public and professional outreach.

National Environmental Trust (NET)
1200 18th Street NW, Fifth Floor
Washington, DC 20036
Telephone: (202) 887-8800
Fax: (202) 887-8877
Web site: http://www.environet.org

As a nonpartisan organization, the NET is dedicated to public education on environmental issues such as air quality and children's health, oil and national security policies, and roadless area protection.

National Ground Water Association (NGWA)
601 Dempsey Road
Westerville, OH 43081
Telephone: (614) 898-7791
Fax: (614) 898-7786
Web site: http://www.ngwa.org

The mission of the NGWA is to provide professional and technical leadership in the advancement of the groundwater industry and to protect, promote, and responsibly develop and use groundwater resources.

National Hispanic Environmental Council
106 North Fayette Street
Alexandria, VA 22314
Telephone: (703) 922-3429

Fax: (703) 922-3761
Web site: http://www.nheec.org

This group works to unite the Hispanic community around issues of the environment and sustainable development, provides an avenue for the Hispanic voice to be heard, and encourages Hispanics to pursue careers in environmental fields.

National Parks Conservation Association
1300 19th Street NW, Suite 300
Washington, DC 20036
Telephone: (800) 628-7275
Fax: (202) 659-0650
Web site: http://www.npca.org

One of the nation's oldest activist groups, the National Parks Conservation Association works to protect national parks throughout the United States through legislative action and grassroots education.

National Recreation and Park Association (NRPA)
22377 Belmont Ridge Road
Ashburn, VA 20148-4501
Telephone: (703) 858-0784
Fax: (703) 858-0794
Web site: http://www.nrpa.org

The organization encourages the advancement of parks, recreation, and environmental conservation efforts. Through public awareness and the dissemination of information, NRPA hopes to increase sound environmental management of parks in the United States.

National Religious Partnership for the Environment (NRPE)
1047 Amsterdam Avenue
New York, NY 10025
Telephone: (212) 316-7441
Fax: (212) 316-7547
Web site: http://www.nrpe.org

A formal alliance of major faith groups and denominations, the NRPE seeks to integrate care for God's creation throughout religious life, including use of public policy initiatives.

National Tree Trust
1120 G Street NW, Suite 770
Washington, DC 20005
Telephone: (202) 628-8733
Fax: (202) 628-8735
Web site: http://www.nationaltreetrust.org

The National Tree Trust acts as a catalyst for local community volunteers who are interested in tree planting and maintenance.

National Tribal Environmental Council (NTEC)
2221 Rio Grande Boulevard NW
Albuquerque, NM 87104
Telephone: (505) 242-2175
Fax: (505) 242-2654
Web site: http://www.ntec.org

NTEC assists tribes in the protection and preservation of reservation environments. It provides technical assistance, training, and information resources to any of America's federally recognized tribes.

National Wilderness Institute
P.O. Box 25766
Washington, DC 20007
Telephone: (703) 836-7404
Fax: (703) 836-7405
Web site: http://www.nwi.org

Calling itself "the voice of reason on the environment," the organization seeks reform of the Endangered Species Act and serves as a litigant in environmental and conservation issues.

National Wildlife Federation (NWF)
11100 Wildlife Center Drive
Reston, VA 20190-5362
Telephone: (703) 438-6000
Fax: (703) 438-3570
Web site: http://www.nwf.org

As one of the nation's oldest environmental organizations, the NWF has a long history of conservation activism on a variety of outdoor issues.

Native American Fish and Wildlife Society
750 Burbank Street
Broomfield, CO 80020
Telephone: (303) 466-1725
Fax: (303) 466-5414
Web site: http://www.nafws.org

The Native American Fish and Wildlife Society was formed to coordinate preservation of Native American fish and wildlife resources.

Native Forest Network
Western North America Office
P.O. Box 8251
Missoula, MT 59807
Telephone: (406) 542-7343
Fax: (406) 542-7347
Web site: http://www.nativeforest.org

The Native Forest Network is a group of forest activists, indigenous peoples, conservation biologists, and nongovernmental organizations. Its goal is to protect the world's remaining native forests.

Natural Resources Defense Council
40 West 20th Street
New York, NY 10011
Telephone: (212) 727-2700
Fax: (212) 727-1773
Web site: http://www.nrdc.org

The National Resources Defense Council works to safeguard wildlife and wild places through legal action, often working in concert with national and international activist groups.

Nature Conservancy
4245 North Fairfax Drive, Suite 100
Arlington, VA 22203-1606
Telephone: (800) 628-6860
Fax: (703) 841-4823
Web site: http://www.nature.org

The Nature Conservancy works with communities, businesses, and people to protect open space in order to preserve plant and animal species.

Northwest Earth Institute
506 SW Sixth Avenue, Suite 1100
Portland, OR 97204
Telephone: (503) 227-2807
Fax: (503) 227-2917
Web site: http://www.nwei.org

Northwest Earth Institute works to motivate people and organizations to help protect the earth. They offer classes to mainstream workplaces to encourage earth-centered conversation.

Pacific Legal Foundation (PLF)
10360 Old Placerville Road, Suite 100
Sacramento, CA 95827
Telephone: (916) 362-2833
Fax: (916) 362-2932
Web site: http://www.pacificlegal.org

"Providing an effective voice in the courts for mainstream thinking" is the goal of the PLF, the largest public interest law firm in the country, focusing on the rights of private property owners in land disputes.

Partners for Livable Communities
1429 21st Street NW
Washington, DC 20036
Telephone: (202) 887-5990
Fax: (202) 466-4845
Web site: http://www.livable.com

This group works with community development institutions, foundations, and city governments to initiate changes designed to increase the livability of communities and is dedicated to increasing and maintaining open spaces in urban areas.

Peninsula Open Space Trust
3000 Sand Hill Road, Building 4, Suite 135
Menlo Park, CA 94025
Telephone: (650) 854-7696
Fax: (650) 854-7703
Web site: http://www.openspacetrust.org

The group works to conserve land on the San Francisco (California) peninsula, from forested hillsides to redwood groves.

People for the Ethical Treatment of Animals (PETA)
501 Front Street
Norfolk, VA 23510
Telephone: (757) 622-7382
Fax: (757) 622-0457
Web site: http://www.peta-online.org

Founded in 1980, PETA operates under the principle that animals are not ours to eat, experiment on, or use for entertainment, educating policymakers and the public about animal abuse and respect.

Pesticide Action Network North America (PANNA)
49 Powell Street, Suite 500
San Francisco, CA 94102
Telephone: (415) 981-1771
Fax: (415) 981-1991
Web site: http://www.panna.org

PANNA works to replace pesticide use with ecologically sound and socially just alternatives as part of one of five Pesticide Action Network regional centers worldwide, linking local and international consumer, labor, health, environment, and agriculture groups.

Physicians for Social Responsibility
1875 Connecticut Avenue NW, Suite 1012
Washington, DC 20009
Telephone: (202) 667-4260
Fax: (202) 667-4201
Web site: http://www.psr.org

Physicians for Social Responsibility works toward the elimination of nuclear weapons and addresses other issues, including global environmental concerns such as global warming.

Planet Drum Foundation
P.O. Box 31251
San Francisco, CA 94131
Shasta Bioregion, USA
Telephone: (415) 285-6556
Fax: (415) 285-6563
Web site: http://www.planetdrum.org

The Planet Drum Foundation formed to encourage community self-sustainability within bioregions. Through education and community action, Planet Drum hopes to encourage other communities to live within their natural confines.

Planning and Conservation League
926 J Street, Suite 612
Sacramento, CA 95814
Telephone: (916) 444-8726
Fax: (916) 448-1789
Web site: http://www.pcl.org

The Planning and Conservation League is a network of concerned citizens and organizations who support sound environmental legislation and administration action.

Political Economy Research Center
502 South 19th Avenue, Suite 211
Bozeman, MT 59718-6827
Telephone: (406) 587-9591
Fax: (406) 586-7555
Web site: http://www.perc.org

This market-oriented think tank focuses on environmental and natural resource issues, including endangered species, fisheries, mines, parks, public lands, property rights, Superfund sites, water, and wildlife.

Population Communication International
777 United Nations Plaza
New York, NY 10017
Telephone: (212) 687-3366
Fax: (212) 661-4188
Web site: http://www.population.org

Population Communication International works with media outlets to make culturally sensitive programming that encourages population choices based on sustainable development and environmental protection.

Public Citizen
1600 20th Street NW

Washington, DC 20009
Telephone: (202) 588-1000
Fax: (202) 547-7392
Web site: http://www.citizen.org

Public Citizen works to have citizens' voices heard in all branches of government. Public Citizen works on issues such as clean energy sources and social justice.

Rails to Trails Conservancy
1100 17th Street NW, Tenth Floor
Washington, DC 20036
Telephone: (202) 331-9696
Fax: (202) 331-9680
Web site: http://www.railtrails.org

The Rails to Trails Conservancy transforms former railroad corridors into public trails ideal for a variety of uses, including walking, cross-country skiing, and bicycling.

Rainforest Action Network
221 Pine Street, Suite 500
San Francisco, CA 94104
Telephone: (415) 398-4404
Fax: (415) 398-2732
Web site: http://www.ran.org

Rainforest Action Network works to protect rain forests worldwide while simultaneously supporting the rights of native inhabitants through programs in education, grassroots organizing, and nonviolent direct action.

Rainforest Alliance
65 Bleeker Street
New York, NY 10012
Telephone: (212) 677-1900
Fax: (212) 677-2187
Web site: http://www.rainforest-alliance.org

Rainforest Alliance seeks to protect tropical rain forests and develop new alternatives to the loss of biodiversity through education, research, and cooperative partnerships.

Redwood Alliance
P.O. Box 293
Arcata, CA 95518
Telephone: (707) 822-7884
Fax: (707) 822-3451
Web site: http://www.redwoodalliance.org

Formed in 1978 by a group of individuals concerned about our nation's growing reliance on nuclear power, the main focus of this grassroots, community-based social and environmental organization is advocacy and education to promote safe and efficient energy use and development.

Reef Relief
P.O. Box 430
Key West, FL 33041
Telephone: (305) 294-3100
Fax: (305) 293-9515
Web site: http://www.reefrelief.org

This membership organization is dedicated to preserving and protecting living coral reef ecosystems through local, regional, and global efforts.

Renewable Energy Policy Project
1612 K Street NW, Suite 410
Washington, DC 20006
Telephone: (202) 293-2833
Fax: (202) 293-5857
Web site: http://www.repp.org

A monthly report on a single aspect of alternative energy is the main activity of this organization, which supports technological development rather than reliance upon subsidies.

Resource Renewal Institute (RRI)
Pier One, Fort Mason Center
San Francisco, CA 94123
Telephone: (415) 928-3774
Fax: (415) 928-6529
Web site: http://www.rri.org

RRI supports innovative environmental management strategies both in the United States and worldwide. Its mission is to catalyze the development and implementation of green plans through projects such as the Water Heritage Trust and the watchdog organization, Defense of Place.

River Network
P.O. Box 8787
Portland, OR 97207
Telephone: (503) 241-3506
Fax: (503) 241-9256
Web site: http://www.rivernetwork.org

This group is dedicated to helping people organize to protect and restore rivers and watersheds at the local level.

Rocky Mountain Institute
1739 Snowmass Creek Road
Snowmass, CO 81654
Telephone: (970) 927-3851
Fax: (970) 927-3420
Web site: http://www.rmi.org

Rocky Mountain Institute is a broad-based institute focusing on many policy areas, including energy and water policy—with all programs grounded in efficient and productive use of resources.

Save Our Wild Salmon (SOWS)
424 Third Avenue West, Suite 100
Seattle, WA 98119
Telephone: (206) 286-4455
Fax: (206) 286-4454
Web site: http://www.sows.org

Save Our Wild Salmon is dedicated to rescuing the Snake River and Columbia River wild salmon from extinction. With field offices around the Northwest and advocates in Washington, D.C., SOWS uses science and public education to promote its position on dams and habitat recovery.

Save-the-Redwoods League
114 Sansome Street, Room 1200
San Francisco, CA 94104-3823
Telephone: (415) 362-2352

Fax: (415) 362-7017
Web site: http://www.savetheredwoods.org

Since 1918, the Save-the-Redwoods League has sought to preserve the redwood forests of California by purchasing redwood
land, which is then turned over to the state or federal park system
or to another public park or reserve.

Scenic America

801 Pennsylvania Avenue SE, Suite 300
Washington, DC 20003
Telephone: (202) 543-6200
Fax: (202) 543-9130
Web site: http://www.scenic.org

Scenic America promotes the scenic heritage of America through
programs such as billboard control, scenic byways, and tree conservation to protect scenic roads and to help people create more
attractive communities.

Sea Shepherd Conservation Society

22774 Pacific Coast Highway
Malibu, CA 90265
Telephone: (310) 456-1141
Fax: (310) 456-2488
Web site: http://www.seashepherd.org

The Sea Shepherd Conservation Society investigates violations of
international laws and helps with enforcement for the protection
of marine wildlife.

Sierra Club

85 Second Street, Second Floor
San Francisco, CA 94105-3441
Telephone: (415) 977-5500
Fax: (415) 977-5799
Web site: http://www.sierraclub.org

The Sierra Club works to protect wild places of the earth, encourage responsible use of resources, and educate citizens about critical environmental issues.

Simple Living Network

P.O. Box 233

Trout Lake, WA 98650
Telephone: (509) 395-2323
Fax: None
Web site: http://www.simpleliving.net

The Simple Living Network offers resources and information for people genuinely interested in living a simpler life.

Society for Conservation Biology
4245 North Fairfax Drive
Arlington, VA 22203
Telephone: (703) 276-2384
Fax: (703) 525-8024
Web site: http://www.conbio.net/scb

The Society for Conservation Biology promotes scientific study in order to maintain, restore, and protect biological diversity.

Society of American Foresters
5400 Grosvenor Lane
Bethesda, MD 20814
Telephone: (301) 897-8720
Fax: (301) 897-3690
Web site: http://www.safnet.org

The society represents the forestry profession and aims to advance the science, education, technology, and practice of forestry. It has significant historical importance to environmental activism because it was founded in 1900 by Gifford Pinchot.

Soil and Water Conservation Society (SWCS)
7515 Northeast Ankeny Road
Ankeny, IA 50021
Telephone: (515) 289-2331
Fax: (515) 289-1227
Web site: http://www.swcs.org

SWCS promotes the science and art of soil, water, and other natural resource management. It works strictly with an ethic recognizing the interdependence of people and the environment.

Southern Alliance for Clean Energy
P.O. Box 1842

Knoxville, TN 37901-1842
Telephone: (865) 637-6055
Fax: (865) 524-4479
Web site: http://www.tngreen.com/cleanenergy

This regional organization focuses on the development of clean air and energy policies, particularly through the reduction of emissions from coal-fired power plants and by promoting alternative forms of clean energy.

Southern Environmental Law Center (SELC)
201 West Main Street, Suite 14
Charlottesville, VA 22902-5065
Telephone: (434) 977-4090
Fax: (434) 977-1483
Web site: http://www.selcga.org

Through the use of litigation and investigation, the SELC works to enforce compliance with existing environmental statutes, often as a part of a coalition with other activist groups.

Southern Utah Wilderness Alliance (SUWA)
1471 South 110 East
Salt Lake City, UT 84105
Telephone: (801) 486-3161
Fax: (801) 486-4233
Web site: http://www.suwa.org

The SUWA works to protect the canyon country and unspoiled public lands surrounding Utah's national parks.

Southern Women Against Toxics (SWAT)
101 West Monroe Street
P.O. Drawer 1526
Livingston, AL 35470
Telephone: (205) 392-7257
Fax: Same as telephone number
Web site: None

SWAT supports women in the role they play within communities to help establish a sense of empowerment, promote self-development, and teach leadership skills on issues of environmental threats in communities due to incinerators and dumps.

Southwest Forest Alliance
P.O. Box 1948
Flagstaff, AZ 86002
Telephone: (928) 774-6514
Fax: (928) 774-6846
Web site: http://www.swfa.org

The coalition of more than fifty environmental groups in Arizona and New Mexico is dedicated to protection of old-growth forests as the best way to protect rural economies, wildlife, ecological diversity, and the interests of future generations.

Stewards of the Range
P.O. Box 1189
Boise, ID 83701
Telephone: (208) 336-5922
Fax: (208) 336-7054
Web site: http://www.stewardsoftherange.org

Stewards of the Range was formed in response to tightening regulations on land use and private property rights and works to roll back restrictive government policies.

Surfrider Foundation
122 South El Camino Real, #67
San Clemente, CA 92672
Telephone: (949) 492-8170
Fax: (949) 492-8142
Web site: http://www.surfrider.org

In 1984, this organization was established by surfers and other citizens who work to protect the oceans, waves, and beaches, through more than fifty chapters in the United States and affiliates in other nations.

Sustainable Ecosystems Institute
0605 Southwest Taylor's Ferry Road
Portland, OR 97219
Telephone: (503) 246-5008
Fax: (503) 246-6905
Web site: http://www.sei.org

Sustainable Ecosystems Institute is a group of scientists who lend their expertise to solve ecological problems.

Taiga Rescue Network
Box 116
SE-96223 Jokkmokk
Sweden
Telephone: 46 971 17039
Fax: 46 971 12057
Web site: http://www.taigarescue.org

This is an international network of nongovernmental organizations, indigenous peoples, and nations working for the protection and sustainable use of the world's boreal forests.

Tides Foundation
P.O. Box 29903
San Francisco, CA 94129
Telephone: (415) 561-6400
Fax: (415) 561-6401
Web site: http://www.tides.org

Since 1976, the group, which is part of a family of organizations, has worked with donors to strengthen community-based non-profit organizations through innovative grant making.

Trout Unlimited
1500 Wilson Boulevard, #310
Arlington, VA 22209-2404
Telephone: (703) 522-0200
Fax: (703) 284-9400
Web site: http://www.tu.org

The group's mission is to conserve, protect, and restore North America's trout and salmon fisheries and their watersheds.

Trust for Public Land
116 New Montgomery Street, Fourth Floor
San Francisco, CA 94105
Telephone: (415) 495-4014
Fax: (415) 495-4103
Web site: http://www.tpl.org

The Trust for Public Land helps to conserve land for recreation through programs devoted to the development of urban parks, watershed protection, and greenways.

Union of Concerned Scientists
2 Brattle Square
Cambridge, MA 02238
Telephone: (617) 547-5552
Fax: (617) 864-9405
Web site: http://www.ucsusa.org

The Union of Concerned Scientists is a group of citizens and scientists who combine rigorous scientific analysis and citizen action to promote campaigns, including fuel-efficient vehicles, renewable energies, and a nuclear-free world.

United States Climate Action Network (USCAN)
1367 Connecticut Avenue NW, Suite 300
Washington, DC 20036
Telephone: (202) 785-8702
Fax: (202) 785-8701
Web site: http://www.climatenetwork.org

USCAN, which serves as the focal point for global climate change activities in the United States, is made up of over thirty-five environment, development, and energy nongovernmental organizations working to inform and affect U.S. domestic and international policies.

United States Public Interest Research Group
State PIRGs
29 Temple Place
Boston, MA 02111
Telephone: (617) 292-4800
Fax: (617) 292-8057
Web site: http://www.pirg.org

The public interest research groups are an alliance of state interest groups devoted to uncovering public health threats, raising money for legislation efforts, and fighting for the public interest.

Urban Land Institute
1025 Thomas Jefferson Street, Suite 500W
Washington, DC 20007
Telephone: (202) 624-7000
Fax: (202) 624-7140
Web site: http://www.uli.org

Publisher of *Land Use Digest* and *Urban Land* magazine, the group works toward improving land use and real estate development.

Water Environment Federation
601 Wythe Street
Alexandria, VA 22314-1994
Telephone: (703) 684-2452
Fax: (703) 684-2492
Web site: http://www.wef.org

Water Environment Federation is a technical and educational organization devoted to the preservation and betterment of the global water supply. Its mission is to promote and advance the water quality industry and to protect and enhance the global water environment.

Water Keeper Alliance
78 North Broadway, E Building
White Plains, NY 10603
Telephone: (914) 422-4410
Fax: (914) 442-4437
Web site: http://www.waterkeeper.org

Water Keeper Alliance is the umbrella organization for more than eighty Water Keeper organizations in North and Central America. The group's goal is the enforcement of environmental regulations and the promotion of community education on local watershed issues.

WaterWatch
213 SW Ash, Suite 208
Portland, OR 97204
Telephone: (503) 295-4039
Fax: (503) 295-2791
Web site: http://www.waterwatch.org

A citizen-based conservation group, WaterWatch monitors the diversion of waters from rivers and streams, often for agricultural purposes.

Wilderness Society
1615 M Street NW
Washington, DC 20036

Telephone: (800) 843-9453
Fax: (202) 429-3957
Web site: http://www.wilderness.org

The Wilderness Society works to protect America's wildlands so that future generations may enjoy clean air and water, wildlife, and recreation opportunities in the wilderness.

WildLaw
300-B Water Street, Suite 208
Montgomery, AL 36104
Telephone: (334) 265-6529
Fax: (334) 265-6511
Web site: http://www.wildlaw.org

This is a nonprofit law firm representing grassroots organizations working to protect the environment, especially on matters dealing with national forests, public lands, endangered species, and protection of waters, wetlands, and coastal areas.

Wildlife Society
5410 Grosvenor Lane
Bethesda, MD 20814
Telephone: (301) 897-9770
Fax: (301) 530-2471
Web site: http://www.wildlife.org

The Wildlife Society augments the ability of wildlife professionals through science and educational programs to ensure continued wildlife stewardship.

Windstar Foundation
P.O. Box 656
Snowmass, CO 81654
Telephone: (970) 927-5435
Fax: Same as telephone number
Web site: http://www.wstar.org

This group is an educational foundation that approaches environmental problems from a holistic premise, recognizing that we are all responsible for and part of the quality of life on earth.

World Rainforest Movement
Maldonado 1858

11200 Montevideo
Uruguay
Telephone: 598 2 413 2989
Fax: 598 2 418 0762
Web site: http://www.wrm.org.uy

This organization publishes a State of the Forests report that analyzes the causes and extent of deforestation, along with recommendations for citizen and political activism.

World Resources Institute
10 G Street NE, Suite 800
Washington, DC 20002
Telephone: (202) 729-7600
Fax: (202) 729-7610
Web site: http://www.wri.org

World Resources Institute is an environmental think tank that seeks practical solutions to environmental issues to help governments and private organizations cope with global change.

World Wildlife Fund
1250 24th Street NW
Washington, DC 20037
Telephone: (202) 293-4800
Fax: (202) 293-9211
Web site: http://www.worldwildlife.org

Dedicated to protecting wildlife and wildlands through addressing issues of endangered spaces and endangered species, the World Wildlife Fund is a prestigious international organization.

Worldwatch Institute
1776 Massachusetts Avenue NW
Washington, DC 20036
Telephone: (202) 452-1999
Fax: (202) 296-7365
Web site: http://www.worldwatch.org

Worldwatch Institute informs the public and policymakers of global environmental concerns and the complex links between the economy and the environment.

Xerces Society
4828 SE Hawthorne Boulevard
Portland, OR 97215
Telephone: (503) 232-6639
Fax: (503) 233-6794
Web site: http://www.xerces.org

The Xerces Society works on protection of invertebrates that are threatened or endangered.

Zero Population Growth
1400 Sixteenth Street NW, Suite 320
Washington, DC 20036
Telephone: (202) 332-2200
Fax: (202) 332-2302
Web site: http://www.zpg.org

Zero Population Growth works to slow population growth in order to achieve a sustainable balance between humans and the earth's resources.

8

Selected Print and Nonprint Resources

Since the advent of the environmental movement in 1970—
Earth Day—there have been thousands of explanations
offered for the rise of grassroots activism, criticisms of that
activism by business and industry interests, coverage of sub-
movements such as environmental justice, explorations of the
philosophical and historical roots of environmentalism, and
speculation on the effectiveness of various political strategies.
This annotated bibliography is designed to provide a road map
to understanding the different forms of activism, the types of
groups that activism has produced, and specific case studies of
individuals and organizations and of the issues around which
they coalesce. There is also a listing of journals and magazines
that regularly cover environmental issues and organizations,
many of which are available on-line or through on-line databases.

Like most maps, the bibliography illustrates the fact that
there are often multiple roads to a single destination, with many
arterials that take longer or are more scenic. Likewise, some of
these books relate engaging stories, other present the results of
academic research and scholarly insight, and some serve as
compendiums of information. Further evidence of the perva-
siveness of activism can be found in the list of electronic databases and web sites that focus exclusively on the environment.
Electronic journals and on-line news sources provide activists
with the latest information at the touch of a few keystrokes on
their computers. In order to get a good sense of what is known

about environmental activism, this chapter provides a choice of options that sometimes overlap or, at other times, contradict one another. But mixed in with the classic books and articles are new ones that raise important issues and ideas—further evidence that environmentalism is still alive and well in the twenty-first century.

Books

Adams, Carol J., ed. 1993. *Ecofeminism and the Sacred.* New York: Continuum. 340 pp.

This edited volume ties together the concept of ecofeminism and religion, providing a feminist perspective on the environment and nature. Various authors identify linkages to the oppression of women, compassion in environmental crisis, the transforming power of symbols, and "ecowomanism."

Arnold, Ron. 1997. *Ecoterror: The Violent Agenda to Save Nature.* Bellevue, WA: Merril Press. 336 pp.

Reissued with updates on the $12 million arson of the Vail ski area and the Unabomber, the book provides a conservative view of the environmental activism of groups such as the Earth Liberation Front and Earth First!

Bari, Judi. 1994. *Timber Wars.* Monroe, ME: Common Courage Press. 343 pp.

Activist Judi Bari of Earth First! reviews the issues surrounding the battles over old-growth forests in northern California in this firsthand account.

Beder, Sharon. 1998. *Global Spin: The Corporate Assault on Environmentalism.* White River Junction, VT: Chelsea Green. 288 pp.

The author explores the role of public relations firms and corporations in "greening" their image and documents various strategies from an international perspective. The book covers the use of environmentally aware names for organizations, suits against activists, and corporate-funded think tanks.

Beierle, Thomas C., and Jerry Cayford. 2001. *Democracy in Practice: Public Participation in Environmental Decisions.* Washington, DC: Resources for the Future Press. 208 pp.

Even in a democratic society, there are questions about how much the public should be involved in environmental decision making. Using various case studies, the researchers identify some of the growing patterns that are now putting participation in greater perspective.

Binnewies, Robert O. 2002. *Palisades.* Bronx, NY: Fordham University Press. 406 pp.

This story of the fight to preserve a unique 100,000-acre interstate system of parks and open spaces on the border of New York and New Jersey began a hundred years ago. It is one of the pioneering efforts in environmental protection that helped shape the modern park conservation movement.

Bliese, John R. E. 2002. *The Greening of Conservative America.* Boulder, CO: Westview Press. 352 pp.

The author defends the case that conservative environmentalism and the growth of a strong economy are not conflicting, arguing that conservative political thought supports the conservation of natural resources and that average Americans can affect the outcome.

Brick, Philip, Donald Snow, and Sarah van de Wetering, eds. 2001. *Across the Great Divide: Explorations in Collaborative Conservation and the American West.* Washington, DC: Island Press. 286 pp.

This is an edited collection of analyses of collaborative conservation and what might be a major change in environmental activism. The various authors deal with case studies such as the Applegate Partnership and watershed efforts as well as accountability in collaboration.

Brulle, Robert J. 2000. *Agency, Democracy, and Nature: The U.S. Environmental Movement from a Critical Theory Perspective.* Cambridge: MIT Press. 496 pp.

Using a broad range of empirical and theoretical research, Brulle develops a methodology for measuring the success of more than

100 U.S. environmental groups. He concludes with an analysis of how groups can make their organizational practices more democratic and politically effective.

Bryant, Bunyan, ed. 1995. *Environmental Justice: Issues, Policies, and Solutions.* Washington, DC: Island Press. 278 pp.

A good starting point for learning the basics about the environmental justice movement, this book provides the ideological framework, the scientific data, and the historical perspective on environmental racism.

Bryner, Gary C. 2001. *Gaia's Wager: Environmental Movements and the Challenge of Sustainability.* Lanham, MD: Rowman and Littlefield. 255 pp.

Tracing the actors, the issues, and the institutions involved in the evolution of environmental groups, the author shows the role that movements play between political parties and interest groups, concluding that only social movements can catalyze a goal of environmental sustainability.

Bullard, Robert D. 2000. *Dumping in Dixie: Race, Class, and Environmental Quality.* 3d ed. Boulder, CO: Westview Press. 256 pp.

Bullard's original groundbreaking work has been updated in its review of the ways in which people of color bear a disproportionate share of the country's environmental problems. This edition provides a fresh perspective on environmental racism, organizing strategies, and success stories.

Bullard, Robert D., ed. 1993. *Confronting Environmental Racism: Voices from the Grassroots.* Boston: South End Press. 259 pp.

This is one of the first comprehensive books on the topic of the environmental justice movement. It includes a number of case studies from communities in Virginia, Alabama, and Colorado, along with activism involving farmworkers, lead poisoning, and industrial exploitation.

Cable, Sherry, and Charles Cable. 1995. *Environmental Problems, Grassroots Solutions.* New York: St. Martin's Press. 143 pp.

The authors credit grassroots environmental activism with a number of successes, from the cleanup and closure of contaminated sites to forcing corporations to consider the environmental impact of their operations, encouraging pollution prevention, and acting as self-help groups.

Camacho, David, ed. 1998. *Environmental Injustices, Political Struggles.* Durham, NC: Duke University Press. 240 pp.

This edited volume provides an overview of the environmental justice movement, offering useful observations on the political strategies that have improved the environmental health and safety of poor minority communities in the United States.

Carle, David. 2002. *Burning Questions: America's Fight with Nature's Fire.* Westport, CT: Praeger. 312 pp.

Although there have been over 100 years of controversy over prescribed burning and fire suppression, there is a tension among environmental activists as to whether or not these policies are useful in forest restoration or just another excuse for cutting down more trees.

Carrels, Peter. 1999. *Uphill against Water: The Great Dakota Water War.* Lincoln: University of Nebraska Press. 238 pp.

Farmers are not usually thought of as environmental activists, but in this case study of a Bureau of Reclamation project in South Dakota, they become an important part of the shaping of water policy in the West.

Cawley, R. McGreggor. 1993. *Federal Land, Western Anger: The Sagebrush Rebellion and Environmental Politics.* Lawrence: University Press of Kansas. 195 pp.

How can a lands dispute in Nevada become such a powerful national issue? The author captures the emotions and arguments of the members of the Sagebrush Rebellion, including cattlemen's associations, wise use advocates, county supremacists, and President Ronald Reagan's responses to the environmental movement.

Chapple, Stephen. 2002. *Confessions of an Eco-Redneck.* Boulder, CO: Perseus Publishing. 264 pp.

Outdoor writer Chapple provides an "alternative" view of environmentalism, with short commentaries on hunting, fishing, and ecology in a witty but persuasive way that is designed to let others know that sportsmen are passionate advocates for the environment.

Cohen, Michael P. 1988. *History of the Sierra Club, 1892–1970.* San Francisco: Sierra Club Books. 550 pp.

As its title indicates, the volume is basically a history of one of the nation's foremost environmental organizations. But what makes it more than just a chronicle is the emphasis on the individuals, such as John Muir, who played such an important role in moving conservation from just an idea to a key political issue.

Cole, Luke W., and Sheila R. Foster. 2001. *From the Ground Up: Environmental Racism and the Rise of the Environmental Justice Movement.* New York: New York University Press. 244 pp.

The authors trace the roots and historical and contemporary causes of environmental racism, considered one of the fastest-growing social movements in the United States. The book examines several well-known case studies of grassroots resistance, including Kettleman City, California; Chester, Pennsylvania; and Dilkon, Arizona.

Coleman, Daniel A. 1994. *Ecopolitics: Building a Green Society.* New Brunswick, NJ: Rutgers University Press. 236 pp.

Writing about the "pillars of effective activism," Coleman provides an analysis of the response to key issues such as population growth, energy strategies, the loss of community, and participatory, grassroots democracy.

Cramer, Phillip F. 1998. *Deep Environmental Politics.* Westport, CT: Praeger. 256 pp.

Cramer argues that radical environmentalists, such as those aligned with the group Earth First!, have played a key role in the shaping of U.S. environmental policy and law, using the tenets of deep ecology as their road map.

Danaher, Kevin, and Roger Burbach. 2000. *Globalize This! The Battle against the World Trade Organization and Corporate Rule.* Monroe, ME: Common Courage Press. 218 pp.

The authors deal with the issues of free trade and international economic relations with developing countries, focusing on the antiglobalization activist organizations. The focus is on the fight against large international corporations.

Davis, Charles, ed. 2001. *Western Public Lands and Environmental Politics.* 2d ed. Boulder, CO: Westview Press. 276 pp.

Political organizations and interest groups have played a significant role in the shaping of the management of rangeland, timber, energy, mineral, and wilderness resources. The eleven articles written for this book explore how activists and political institutions have affected policies through the Clinton administration.

Dean, Cornelia. 1999. *Against the Tide: The Battle for America's Beaches.* New York: Columbia University Press. 279 pp.

This book investigates the nature of the East, West, and Gulf Coasts of the United States and presents stories of development and destruction by hurricanes that have resulted in ailing coastlines. It explores the human element as policymakers make an attempt to decide what to do about these ecosystems where humans have become intruders.

DeLuca, Kevin Michael. 1999. *Image Politics: The New Rhetoric of Environmental Activism.* New York: Guilford Press. 205 pp.

Using the language and context of environmental activism, the author looks at the way social movements create and maintain an image to effect social change.

Donahue, Debra L. 1999. *The Western Range Revisited.* Norman: University of Oklahoma Press. 388 pp.

The landscape of the West is captured in this book's emphasis on the proposed removal of livestock from public lands to conserve native diversity. It presents a well-researched perspective on the role of ranchers as political activists, the evangelical environmental movement, and biodiversity advocates.

Duncan, Glen A. 2000. *Goodbye Green: How Extremists Stole the Environmental Movement from Moderate America and Killed It.* Bellevue, WA: Merril Press. 182 pp.

"A dying movement" is how the author characterizes environmentalism in the United States. He notes that the lack of moderate voices has scared away those who might otherwise be engaged, along with biased media attention. The chemical corridor of south Louisiana is examined, along with the role of groups such as the Audubon Society and the Sierra Club.

Duncan, James R., and John E. Kinney. 1997. *Conservative Environmentalism.* Westport, CT: Quorum Books. 296 pp.

Asset production is the focus of this tightly documented book, the theme of which is that the environmental movement is a negative, rather than positive, group of activists who are not really helping to solve environmental problems.

Engberg, Robert, and Donald Wesling, eds. 1999. *John Muir: To Yosemite and Beyond.* Salt Lake City: University of Utah Press. 171 pp.

This "composite autobiography" is culled from the letters, journals, articles, and unpublished manuscripts of John Muir, perhaps one of the nation's most important original environmental activists.

Foreman, Dave. 1991. *Confessions of an Eco-Warrior.* New York: Harmony Books. 228 pp.

As one of the leaders of Earth First!, Foreman uses this book as a way of explaining his involvement in the radical activist group that refuses to be classified as an "organization."

Fox, Jonathan A., and L. David Brown, eds. 1998. *The Struggle for Accountability: The World Bank, NGOs, and Grassroots Movements.* Cambridge: MIT Press. 570 pp.

This lengthy, edited volume explores the role of the World Bank and environmental reform, using case studies of activism in specific countries and advocacy campaigns. Grassroots efforts are perceived to be an effective way of countering corporate interests.

Fox, Stephen R. 1985. *John Muir and His Legacy: The American Conservation Movement.* Madison: University of Wisconsin Press. 436 pp.

This is a reprint of a book originally published in 1981 by Little, Brown that is more than just biographical information. Fox expands the coverage to show the impact Muir has had on the modern environmental movement and how his ideas and beliefs continue to influence activists.

Freeman, Jo, and Victoria Johnson, eds. 1999. *Waves of Protest: Social Movements since the Sixties.* Lanham, MD: Rowman and Littlefield. 381 pp.

From the origins of social movements to recruitment, organization, and mobilization, this book uses the work of numerous researchers and case studies to explain how groups form and grow.

Goodman, Doug, and Daniel McCool, eds. 1999. *Contested Landscape: The Politics of Wilderness in Utah and the West.* Salt Lake City: University of Utah Press. 266 pp.

This is a unique book because it grew out of a political science course taught at the University of Utah. Students were given an assignment of writing or cowriting a chapter based on their original research on wilderness policy in the West.

Gottlieb, Allan, ed. 1989. *The Wise Use Agenda: The Citizen's Policy Guide to Environmental Resource Issues.* Bellevue, WA: Free Enterprise Press. 167 pp.

Subtitled *The Citizen's Policy Guide to Environmental Resource Issues,* this book is the outcome of the Multiple Use Strategy Conference held in August 1988 in Reno, Nevada. The event was sponsored by the Center for the Defense of Free Enterprise, and the book argues for changes in the way the environment is managed to meet the needs "of all Americans."

Gottlieb, Robert. 2001. *Environmentalism Unbound.* Cambridge: MIT Press. 408 pp.

This critique of the environmental movement is based on the assumption that activists have been isolated from the vital issues of everyday life because of their narrow interpretation of what is meant by "the environment." The author uses three case studies to show how a broader agenda that includes livable communities will bridge the gap with the natural world.

Gould, Kenneth. 1996. *Local Environmental Struggles: Citizen Activism in the Treadmill of Production.* New York: Cambridge University Press. 239 pp.

Case studies of communities facing uncertainty due to technological change are the focus of Gould's book, which includes coverage of community power and regional planning.

Graf, William L. 1990. *Wilderness Preservation and the Sagebrush Rebellions.* Lanham, MD: Rowman and Littlefield. 329 pp.

Graf is one of the few scholars to distinguish among the various forms of western activism that have become collectively known as the Sagebrush Rebellion. He argues that there is a long history of attempts to preserve wilderness areas through various management strategies.

Hays, Samuel P. 1959. *Conservation and the Gospel of Efficiency: The Progressive Conservation Movement, 1890–1920.* Cambridge: Harvard University Press. 297 pp.

In this pioneering volume, Hays examines the period from 1890 to 1920 and the beginnings of what would become the environmental ethic in the United States.

Hays, Samuel P. 1987. *Beauty, Health, and Permanence: Environmental Politics in the United States, 1955–1985.* New York: Cambridge University Press. 630 pp.

Considered one of the foremost environmental historians in the United States, Hays provides a comprehensive history of the development of the activist movement from 1955 to 1985, exploring the societal and cultural factors that led to contemporary environmentalism.

Helvarg, David. 1994. *The War against the Greens: The "Wise Use" Movement, the New Right, and Anti-Environmental Violence.* San Francisco: Sierra Club Books. 502 pp.

This book traces American attitudes toward the environment and the rise of the wise use movement during the Reagan-Bush administrations. It looks to motives, theories, and the tactics of antienvironmental groups.

Hofrichter, Richard, ed. 2000. *Reclaiming the Environmental Debate: The Politics of Health in a Toxic Culture.* Cambridge: MIT Press. 356 pp.

The seventeen essays in this book cover a broad range of toxic ills, from cancer and brownfields to worker health and mining companies. The book also covers issues related to environmental justice and the ways in which various communities have responded to toxic threats.

Huber, Peter W. 1999. *Hard Green: Saving the Environment from the Environmentalists: A Conservative Manifesto.* New York: Basic Books. 256 pp.

This new environmental agenda, dubbed Hard Green, criticizes contemporary activism because, the author argues, it actually works to destroy the natural environment. The new manifesto should involve those who rediscover Theodore Roosevelt, the conservationist ethic, and a reversal of what Huber calls the Soft Green agenda.

John, DeWitt. 1994. *Civic Environmentalism.* Washington, DC: Congressional Quarterly Press. 341 pp.

In the 1980s, considerable attention was given to the idea that instead of forcing state and local governments to accept "top-down regulation," communities should organize on their own to protect the environment. The concept includes such proposals as curbside recycling, economic incentives, and public education developed by citizens in their own neighborhoods.

Kalt, Brian C. 2001. *Sixties Sandstorm: The Fight over Establishment of a Sleeping Bear Dunes National Lakeshore, 1961–1970.* East Lansing: Michigan State University Press. 151 pp.

This case study is illustrative of the political, social, and cultural issues that sometimes clash—this time over the creation of a national lakeshore in lower Michigan. It explores how citizens organized themselves, how they dealt with the political process, and how this took place just as the environmental movement was going through a period of major change and growth.

Kassman, Kenn. 1998. *Envisioning Ecotopia: The U.S. Green Movement and the Politics of Radical Social Change.* Westport, CT: Praeger. 160 pp.

This slim volume stems from Kassman's experiences as an environmental activist. He distinguishes among three elements of the green movement: neoprimitivism, mystical deep ecology, and social ecology. These subcultures involve differing worldviews and visions for the future, he notes, and have affected the more radical members of the movement.

Kidner, David W. 2001. *Nature and Psyche: Radical Environmentalism and the Politics of Subjectivity*. Albany: State University of New York Press. 320 pp.

Radical environmentalists are trapped in a paradigm that involves a simplistic opposition to technology, Kidner says, and a more critical historical and cultural awareness would increase participation and increase cultural diversity.

Kirk, Andrew Glenn. 2002. *Collecting Nature: The American Environmental Movement and the Conservation Library*. Lawrence: University Press of Kansas. 256 pp.

In 1960, Denver established its Conservation Library as a repository for environmental and conservation documents. The library is "a microcosm of the movement itself" and a clear barometer of the shifts in gender and generations of environmental activists. It tracks the evolution of activism from conservationist ideology to environmentalism.

Kirkman, Robert. 2002. *Skeptical Environmentalism: The Limits of Philosophy and Science*. Bloomington: Indiana University Press. 224 pp.

By using the writings of theorists such as Immanuel Kant, Georg W. F. Hegel, René Descartes, Jean-Jacques Rousseau, and Martin Heidegger, the author shows how our knowledge of nature is developed, the establishment of an environmental worldview, and the promise of a more practical attitude that balances conflicts between values and worldviews.

Knight, Richard L., Wendell C. Gilgert, and Ed Marston. 2002. *Ranching West of the 100th Meridian: Culture, Ecology and Economics*. Washington, DC: Island Press. 259 pp.

The contributors to this volume begin by bringing to life the culture of livestock ranching in the West, which has been attacked by environmentalists and fiscal conservatives. The second section deals with the ecology of ranching, followed by the impact of declining commodity prices and rising land values brought on by the suburbanization of the West.

Lester, James P., David Allen, and Kelly M. Hill. 2001. *Environmental Injustice in the U.S.: Myths and Realities.* Boulder, CO: Westview Press. 232 pp.

Beginning with the history of the environmental justice movement, the authors not only look at scholarly literature but also provide a quantitative analysis of the relationship of race, class, political mobilization, and environmental harm at the state, county, and city levels. They conclude that race and class are significant variables in explaining patterns of environmental harm.

Libby, Ronald T. 1998. *Eco-Wars.* New York: Columbia University Press. 254 pp.

Exploring the strategies of what he calls "expressive interest groups," Libby reviews the success of activists within several environmental campaigns: antibiotechnology, animal rights, California's Big Green initiative, endangered species, and secondhand smoke.

List, Peter C. 1993. *Radical Environmentalism: Philosophy and Tactics.* Belmont, CA: Wadsworth Publishing. 276 pp.

This edited volume is noteworthy because the anthology covers deep ecology, ecofeminism, and social ecology while profiling ecotactics by groups such as Greenpeace, Earth First!, and the Sea Shepherd Society.

Mason, Michael. 1999. *Environmental Democracy.* New York: St. Martin's Press. 266 pp.

The author looks at the environmental implications of a discourse theory of democracy—a model that refers to communications, public participation, and political outcomes. He uses case studies to illustrate the role of discourse, including the countries of the European Union, forest planning in British Columbia, and wilderness values and deep ecology.

McDonald, David A., ed. 2002. *Environmental Justice in South Africa.* Athens: Ohio University Press. 312 pp.

Although the issue of environmental racism has most often been analyzed in an American context, this edited volume provides a series of environmental justice theory and practice stories with essays from South Africa, exploring the struggles of workers and communities seeking environmental change.

McKean, Margaret. 1981. *Environmental Protest and Citizen Politics in Japan.* Berkeley: University of California Press. 291 pp.

Citizen participation in Japan is markedly different from that in the United States, and environmental activism is relatively new and often modeled after protests and demonstrations in other nations.

Mercier, Jean. 1998. *Downstream and Upstream Ecologists: The People, Organizations, and Ideas behind the Movement.* Westport, CT: Praeger. 240 pp.

Ecology has become an overused word in environmentalism, but the author attempts to put the idea into perspective in this background volume.

Meyer, John M. 2001. *Political Nature: Environmentalism and the Interpretation of Western Thought.* Cambridge: MIT Press. 224 pp.

By looking at contemporary environmentalist thinkers, the author concludes that we must expand our thinking about the western portrayals of nature and politics. He uses case studies of struggles over toxic waste dumps in poor neighborhoods, land use in the American West, and rain forest protection in the Amazon.

Miller, Char. 2001. *Gifford Pinchot and the Making of Modern Environmentalism.* Washington, DC: Island Press. 458 pp.

This lengthy and well-documented biography retells the life of one of the nation's most important environmental leaders. It includes extensive information about the struggle between Pinchot and John Muir, based on correspondence and personal papers, in a very personal sketch of Pinchot.

Miller, Norman. 2002. *Environmental Politics: Interest Groups, the Media, and the Making of Policy.* Boca Raton, FL: Lewis Publishers. 174 pp.

Miller outlines what he terms "the shifting tides of environmental advocacy" and the new emphasis of on-line activism from his perspective as a "foot soldier in the environmental wars." Although short, the book does provide an extensive analysis of how technology is shaping policymaking and how the interplay of interest groups, think tanks, and other actors affects the legislative and regulatory processes.

Mitchell, John Hanson. 1999. *Trespassing: An Inquiry into the Private Ownership of Land.* Cambridge, MA: Perseus Publishing. 320 pp.

Property rights activists represent a vocal segment of the environmental opposition, and Mitchell tells their story, from British common law to the most recent U.S. Supreme Court decisions on "takings."

Mongillo, John, and Bibi Booth, eds. 2001. *Environmental Activists.* Westport, CT: Greenwood Press. 368 pp.

This book of profiles identifies over eighty environmental activists from various segments of the movement, putting a human face on very complicated and emotional debates.

Nichols, John T. 2001. *An American Child Supreme: The Education of a Liberation Ecologist.* Minneapolis, MN: Milkweed Editions. 197 pp.

The author of *The Milagro Beanfield War* travels to Guatemala in a journey that leads him to environmental and social activism after a life of privilege.

Novotny, Patrick. 2000. *Where We Live, Work, and Play: The Environmental Justice Movement and the Struggle for Environmentalism.* Westport, CT: Praeger. 136 pp.

This basic and fundamental look at the environmental justice movement examines the growing awareness of environmental problems in minority communities throughout the United States.

Pulido, Laura. 1996. *Environmentalism and Economic Justice: Two Chicano Struggles in the Southwest.* Tucson: University of Arizona Press. 282 pp.

Analyzing "subaltern environmental struggles," the author examines the United Farm Workers pesticide campaign from 1965 to 1971 and the Ganados del valle activists.

Reiger, John F. 1975. *American Sportsmen and the Origins of Conservation.* New York: Winchester Press. 316 pp.

Sometimes hunters and anglers are only interested in the kill and the catch, but this book shows the history of sportsmen's conservation actions that have spearheaded the creation of national parks, forests, and wildlife refuges.

Rogers, Raymond Albert. 1998. *Solving History: The Challenge of Environmental Activism.* Montreal: Black Rose. 211 pp.

This book combines a thorough overview of environmental protection and activism, including an analysis of the environmental aspects of economic development and human ecology.

Rome, Adam. 2001. *The Bulldozer in the Countryside: Suburban Sprawl and the Rise of American Environmentalism.* New York: Cambridge University Press. 299 pp.

Covering the period of suburban growth from 1945 to 1970, the author looks at the mass migration of Americans to the suburbs and the development of the environmental movement. He documents how environmental problems became part of the agenda and how groups responded.

Rose, Fred. 1999. *Coalitions across the Class Divide: Lessons from the Labor, Peace, and Environmental Movements.* Ithaca, NY: Cornell University Press. 272 pp.

Only by bridging working- and middle-class movements and cultures can the struggle for jobs and economic justice mesh with the goals of the environmental movement. Rose argues for a coalition that brings together a series of case studies that explores the practical lessons of organizing across movements and classes.

Rothman, Hal K. 1998. *The Greening of a Nation? Environmentalism in the United States since 1945.* Fort Worth, TX: Harcourt Brace. 219 pp.

By exploring the period since World War II, Rothman explains how environmentalism has become a fixture of U.S. culture. He uses the case study of the Echo Park Dam controversy, the back-to-nature movement of the late 1960s, environmental justice, and the Sagebrush Rebellion as examples of the changes within activism.

Rowell, Andrew. 1996. *Green Backlash: Global Subversion of the Environmental Movement.* London: Routledge. 504 pp.

This account of what Rowell terms the "anti-environmental movement" gives global coverage to groups that have emerged in the last two decades. It chronicles the forestry debates in Canada and Australia as well as conflicts over marine resources in Europe and Southeast Asia and shows how violence and threats have suppressed activism throughout the world.

Rubin, Charles T. 1994. *The Green Crusade: Rethinking the Roots of Environmentalism.* New York: Free Press. 312 pp.

Rubin describes the seminal personalities in American conservation and science, including Rachel Carson, Paul Ehrlich, and the Club of Rome, from a very critical perspective.

Salazar, Debra J., and Donald K. Alper, eds. 2001. *Sustaining the Forests of the Pacific Coast: Forging Truces in the War in the Woods.* Vancouver: University of British Columbia Press. 264 pp.

This collection of essays offers new perspectives on the complexity of the issues in U.S. and Canadian forest policy. It looks at the conflicts of the last twenty years that have polarized communities and activists on both sides of the border, including members of First Nations.

Schneider, Richard J., ed. 2000. *Thoreau's Sense of Place.* Iowa City: University of Iowa Press. 310 pp.

This collection of essays provides an in-depth look at the philosopher and his impact on the environmental movement. It includes studies that compare Thoreau's work with other leaders in environmental philosophy and covers his views on environmental protection.

Schwab, Jim. 1994. *Deeper Shades of Green: The Rise of Blue-Collar and Minority Environmentalism in America*. San Francisco: Sierra Club Books. 490 pp.

One of the few books that explores environmental activism as a phenomenon outside the sphere of white elites, this volume makes a contribution to the literature, especially as it relates to labor interests.

Seel, Benjamin, Matthew Paterson, and Brian Doherty. 2000. *Direct Action in British Environmentalism*. London: Routledge. 223 pp.

The radical environmental movement in Great Britain began using protests and demonstrations in the early 1990s, focused on genetically engineered foods, the car culture and road building, and rampant consumerism. The use of militant tactics is discussed in depth, along with their impact.

Shabecoff, Philip. 1993. *A Fierce Green Fire: The American Environmental Movement*. New York: Hill and Wang. 352 pp.

Investigating how resource management decisions are handled politically, Shabecoff highlights the presidential legacies of Franklin D. Roosevelt and Ronald Reagan, among others. This is a classic overview of American environmentalism written from a journalist's perspective.

Shabecoff, Philip. 2001. *Earth Rising: American Environmentalism in the 21st Century*. Washington, DC: Island Press. 258 pp.

As a freelance writer, Shabecoff's works have brought a unique perspective to the study of American environmentalism. Based on an extensive range of interviews, he considers the need for a broader, more inclusive environmentalism that looks at issues such as education reform, the global economy, and the development of a moral center.

Shaw, Randy. 1999. *Reclaiming America: Nike, Clean Air, and the New National Activism*. Berkeley: University of California Press. 312 pp.

Shaw believes that the environmental movement in the United States has broadened in the past decade to include new areas of activism, such as the antisweatshop campaigns and protests against consumerism. He outlines what is happening with community-based organizations, the media, and the Internet as resources for activists.

Shutkin, William A. 2000. *The Land That Could Be: Environmentalism and Democracy in the Twenty-First Century.* Cambridge: MIT Press. 340 pp.

According to the author, twenty-first-century environmental activism is expanding to include civic health and sustainable local economies. This "new" activism is based on the idea that environmentalism "is as much about protecting ordinary places as it is about preserving wilderness areas."

Steel, Brent S., ed. 1997. *Public Lands Management in the West: Citizens, Interest Groups, and Values.* Westport, CT: Praeger. 224 pp.

This interdisciplinary examination of the public lands debate includes an analysis of the role of public values and philosophical views about the environment, along with the ways in which citizen activists can influence policymaking.

Stern, Alissa J., and Tim Hicks. 2000. *The Process of Business/Environmental Collaboration.* Westport, CT: Quorum Books. 224 pp.

Written in a positive tone, the authors believe that there are many ways in which business and environmental organizations can resolve their differences and views to produce outcomes that are valuable to both.

Sussman, Glen. 2002. *American Politics and the Environment.* New York: Longman. 334 pp.

This basic primer on environmental policy includes an analysis of the role of public opinion, political parties, and interest groups and the role of these actors in dealing with environmental problems.

Sutton, Philip W. 2000. *Explaining Environmentalism: In Search of a New Social Movement.* Aldershot, Hampshire, UK: Ashgate. 248 pp.

By connecting environmentalism to social movement theory, Sutton examines the rise of the environmental movements in the United States and Great Britain. Historical and sociological perspectives on environmental activism are woven together, along with theories about radical ecology and nature politics.

Switzer, Jacqueline Vaughn. 1997. *Green Backlash: The History and Politics of Environmental Opposition in the U.S.* Boulder, CO: Lynne Rienner. 323 pp.

In arguing that there are three separate segments of grassroots environmental opposition (wise use, county supremacy, and private property groups), the author distinguishes between industry opposition and citizen opposition. The book includes discussion of the strategies used by each segment and analyzes their impact on environmental policy.

Tesh, Sylvia Noble. 2001. *Uncertain Hazards: Environmental Activists and Scientific Proof.* Ithaca, NY: Cornell University Press. 192 pp.

Disputes over possible linkages between public health and pollution are often carried on with little scientific evidence. Using community activists' experiences, Tesh notes that many of the debates are often carried out by two different sets of "experts" with culturally important moral dimensions.

Turner, Frederick W. 2001. *John Muir: Rediscovering America.* Cambridge, MA: Perseus Publishing. 448 pp.

This biography of John Muir is notable because of the reputation of its author, a noted historian, and because it covers Muir's life from boyhood, to an almost solitary existence, to a trusted friend of Theodore Roosevelt.

Vale, Thomas R., ed. 2002. *Fire, Native Peoples, and the Natural Landscape.* Washington, DC: Island Press. 238 pp.

This book attempts to answer the question of how Native Americans modified the early landscape of pre-European America. The seven chapters, each by a different author, deal with different subregions of the West and the "naturalness" of these areas, which some environmental activists are now attempting to "restore" to a pristine state.

Wall, Derek. *Earth First! and the Anti-Roads Movement.* 1999. New York: Routledge. 219 pp.

One of the few things the United States has exported to Great Britain is ecoterrorism, used in Great Britain to stop the building of new roads or the expansion of roadways in scenic areas.

Walsh, Edward J., Rex Warland, and D. Clayton Smith. 1997. *Don't Burn It Here: Grassroots Challenges to Trash Incinerators.* University Park: Pennsylvania State University Press. 292 pp.

The siting of waste incinerators has been an issue that has galvanized grassroots environmental activists and led to numerous studies of democratic decision making. This study of municipal solid waste projects in New York, New Jersey, and Pennsylvania explores the actions of ordinary citizens fighting local proposals to burn waste.

Wapner, Paul K. 1996. *Environmental Activism and World Civic Politics.* Albany: State University of New York Press. 238 pp.

Three nongovernmental organizations are highlighted in this book: Greenpeace, the World Wildlife Fund, and Friends of the Earth. The analyses explore their role in global environmental issues and the role of the state.

Warren, Karen J., and Duane L. Cady, eds. 1996. *Bringing Peace Home: Feminism, Violence and Nature.* Bloomington: Indiana University Press. 235 pp.

This is a scholarly work on ecofeminism, an anthology of sixteen authors and fourteen chapters. It developed from the editor's course on Peace, Ecology, and Feminism at Hamline University. The purpose of the book is to encourage and expand scholarship that connects feminist philosophy and peace scholarship.

Warren, Louis S. 1997. *The Hunter's Game: Poachers and Conservationists in Twentieth-Century America.* New Haven: Yale University Press. 227 pp.

This book uses game law enforcement in three states—Pennsylvania, New Mexico, and Montana—to investigate the broader issues of class, power, and nature. The author argues that conservationists were actually hunters whose interests were driven by recreation and tourism as early as the late nineteenth century.

Wellock, Thomas R. 1998. *Critical Masses: Opposition to Nuclear Power in California, 1958–1978.* Madison: University of Wisconsin Press. 333 pp.

Although Californians have always enjoyed a reputation as being on the cutting edge of whatever trends are about to sweep the rest of the nation, this book explores how this was especially true in the state before the environmental movement coalesced. By looking at the efforts to oppose the building of nuclear power plants, Wellock reveals the deep sense of responsibility citizens felt toward protection of their state.

Weltzien, O. Alan, ed. 2002. *The Literary Art and Activism of Rick Bass.* Salt Lake City: University of Utah Press. 318 pp.

Fifteen scholars analyze the work of Rick Bass, an activist, nature writer, and ecojournalist who questioned how writers could engage in the celebration of the earth in the face of environmental degradation.

Western, David. 1997. *In the Dust of Kilimanjaro.* Washington, DC: Island Press. 297 pp.

One of Kenya's most prominent citizens recounts his efforts to protect the nation's wildlife. In addition to offering a chronicle of the history of African wildlife conservation, he explores the political and cultural struggles he has faced.

Wheat, Frank. 1999. *California Desert Miracle: The Fight for Desert Parks and Wilderness.* San Diego, CA: Sunbelt Publications. 367 pp.

This is the story of how underpaid and underfunded volunteers fought for the last large area of wildland left in America, which culminated in the enactment of the California Desert Protection Act of 1994.

Williams, Walter Lee, Jr. 2002. *Determining Our Environments: The Role of Energy Citizen Advisory Boards.* Westport, CT: Praeger. 190 pp.

At a time when energy issues and the cost of power are on the mind of every citizen, Williams looks at how public agencies can better incorporate citizen participation in the decision-making process.

Wood, Paul. 2000. *Biodiversity and Democracy: Rethinking Society and Nature.* Vancouver: University of British Columbia Press. 254 pp.

The author argues that the problem of extinction can be traced to the ways humans think about democratic society and biodiversity. He sees the issue as one of intergenerational justice and notes that biodiversity should be conserved even if it is not in the public's best interest to do so.

Reference Books and Directories

Berry, John, and E. Gene Frankland, eds. 2002. *International Encyclopedia of Environmental Politics.* New York: Routledge. 560 pp.

There are over 500 entries in this volume, which is also cross-referenced for further reading. Sections include prominent individuals who have been actively involved in international environmental issues, organizations, and struggles on the local to international levels.

Callahan, John. *Environmental Directory: Organizations from around the Globe.* http://www.earthrights.com.

This is an on-line directory that includes organizations, trusts, associations, departments, and individuals. It is produced by an activist from New Zealand.

Clark, Jeanne Nienaber, and Hanna J. Cortner. 2002. *The State and Nature: Voices Heard, Voices Unheard in America's Environmental Dialogue.* Upper Saddle River, NJ: Prentice-Hall. 357 pp.

One of the best ways to understand environmental activism is to examine primary documents that can be correlated with specific political actions. By providing a historical dimension to activism, the forty-four selections cover the period 1780 to 2001 and a diverse set of issues.

Harbinger Communications. *The National Environmental Directory.* http://www.environmentaldirectory.net.

This privately produced directory started in 1978 and is now available on-line and as a software program. It includes more than 13,000 organizations in the United States concerned with environmental issues and environmental education. Organizations can add and update their own listings to make the site current.

National Wildlife Federation. 2002. *2002 Conservation Directory.* New York: Lyons Press. 700 pp.

The national organization has compiled an exhaustive listing of organizations dealing with conservation issues, and the directory is updated regularly.

Neimark, Peninah, and Peter Rhoades Mott, eds. 1999. *The Environmental Debate: A Documentary History.* Westport, CT: Greenwood Press. 319 pp.

A comprehensive analysis of environmental activism, divided into six major historical periods from 1776 to 2000. The authors use primary documents to show how speeches, statutes, and statistics have influenced environmental policy. The book includes an extensive bibliographic section and index.

Patton-Hulce, Vicki R. 1995. *Environment and the Law: A Dictionary.* Santa Barbara, CA: ABC-CLIO. 361 pp.

This reference work provides a summary of legal cases involving environmental groups as well as explanations of their importance.

Journals and Magazines

Note that some of these publications are now available on-line, whereas others are available through subscription or through membership in an organization. Some also publish only their table of contents on-line, although individual articles can sometimes be accessed through the electronic databases listed later in this chapter.

American Naturalist
The University of Chicago Press
1427 East 60th Street
Chicago, IL 60637
Telephone: (773) 702-0446
Web site: http://www.journals.uchicago.edu/AN/home.html

Arctic
Arctic Institute of North America
The University of Calgary
2500 University Drive NW
Calgary, Alberta
Canada T2N 1N4
Telephone: (403) 220-4035
Web site: http://www.ucalgary.ca/aina/pubs/members

Audubon
National Audubon Society
700 Broadway
New York, NY 10003
Telephone: (212) 979-3000
Web site: http://www.magazine.audubon.org/

The Auk
Department of Biological Sciences
WAAX 19
University of Arkansas
Fayetteville, AK 72701
Telephone: (501) 575-4683
Web site: http://www.aou.org/aou/auklet.html

BioCycle
The JG Press, Inc.
419 State Avenue
Emmaus PA, 18049
Telephone: (610) 967-4135, ext. 22
Web site: http://www.jgpress.com/biocycle.htm

Conservation Biology
Blackwell Science Limited
Osney Mead, Oxford
United Kingdom OX2 0EL

Telephone: +44 1865 206206
Web site: http://www.blackwell-science.com

Coral Reefs
Springe-Verlag New York, Inc.
P.O. Box 2485
Secaucus, NJ 07096-2485
Telephone: (800) SPRINGER
Web site: http://link.springer.de/link/service/journals

Ducks Unlimited
Ducks Unlimited, Inc.
One Waterfowl Way
Memphis, TN 38120
Telephone: (901) 758-3825
Web site: http://www.ducks.org/media/magazine/

E—The Environmental Magazine
28 Knight Street
Norwalk, CT 06851
P.O. Box 5098
Westport, CT 06881
Telephone: (203) 854-5559
Web site: http://www.emagazine.com/

Earth Island Journal
300 Broadway, Suite 28
San Francisco, CA 94133-3312
Telephone: (415) 788-3666
Web site: http://www.earthisland.org/eijournal/journal.cfm

Ecologist
The Ecologist
P.O. Box 326
Sittingbourne, Kent
United Kingdom ME9 8FA
Telephone: +44 (0) 1795 414963
Web site: http://www.theecologist.org/

Ecos
Csiro Publishing
P.O. Box 1139

Collingwood, Victoria
Australia 3066
Telephone: +61 3 9662 7666
Web site: http://www.publish.csiro.au/ecos/

Environment
Heldref Publications
1319 Eighteenth Street NW
Washington, DC 20036-1802
Telephone: (800) 365-9753
Web site: http://www.heldref.org

Environment International
Elsevier Science
P.O. Box 945
New York, NY 10159-0945
Telephone: (212) 633-3730
Web site: http://www.elsevier.nl/inca/publications

Environmental Forum
Environmental Law Institute
1616 P Street NW, Suite 200
Washington, DC 20036
Telephone: (202) 939-3800
Web site: http://www.eli.org

Environmental History
701 Wm. Vickers Avenue
Durham, NC 27701
Telephone: (919) 682-9319
Web site: http://www.lib.duke.edu/forest/ehmain

Environmental Protection
Stevens Publishing Corporation
5151 Beltline Road, Tenth Floor
Dallas, TX 75254
Telephone: (972) 687-6700
Web site: http://www.stevenspublishing.com/ep

Environmental Reviews
NRC Research Press
National Research Council of Canada

Ottawa, Ontario
Canada K1A 0R6
Telephone: (613) 993-0362
Web site: http://www.nrc.ca/cisti/journals/rp2_home

Environmental Values
The White Horse Press
10 High Street
Knapwell, Cambridge
United Kingdom CB3 8NR
Telephone: +44 1954 267527
Web site: http://www.erica.demon.co.uk/EV

Environmentalist
Kluwer Academic Publishers
P.O. Box 358, Accord Station
Hingham, MA 02018-0358
Telephone: (781) 871-6600
Web site: http://www.kluweronline.com

Everyone's Backyard
Center for Health, Environment, and Justice
P.O. Box 6806
Falls Church, VA 22040
Telephone: (703) 237-2249
Web site: http://www.chej.org/backyard.html

Fisheries
American Fisheries Society
5410 Grosvenor Lane, Suite 110
Bethesda, MD 20814
Telephone: (301) 897-8616
Web site: http://www.fisheries.org/fisheries

Forest Magazine
Forest Service Employees for Environmental Ethics
P.O. Box 11615
Eugene, OR 97440
Telephone: (541) 484-2692
Web site: http://www.fseee.org

Forestry Chronicle
The Forestry Chronicle
151 Slater Street, Suite 606
Ottawa, Ontario
Canada K1P 5H3
Telephone: (613) 234-2242
Web site: http://www.cif-ifc.org/chron.html

Forum for Applied Research and Public Policy
University of Tennessee, EERC
311 Conference Building
Knoxville, TN 37996-4134
Telephone: (865) 974-4251
Web site: http://forum.ra.utk.edu/

High Country News
119 Grand Avenue
P.O. Box 1090
Paonia, CO 81428
Telephone: (970) 527-4898
Web site: http://www.hcn.org

International Journal of Global Environmental Issues
Inderscience Enterprises Limited
World Trade Center Building
29, route de Pre-Bois
Case Postale 896
CH - 1215 Geneva 15
Switzerland
Telephone: ++44 1234 240519
Web site: http://www.inderscience.com/catalogue/g/ijgenvi/indexijgenvi.html

Journal of Forestry
SAF Sales
5400 Grosvenor Lane
Bethesda, MD 20814-2198
Telephone: (301) 897-8720, ext. 106
Web site: http://www.safnet.org/pubs/periodicals

National Parks
National Parks Conservation Association

1300 19th Street NW, Suite 300
Washington, DC 20036
Telephone: (800) 628-7275
Web site: http://www.npca.org

National Resources Forum
Elsevier Science
P.O. Box 945
New York, NY 10159-0945
Telephone: (212) 633-3730
Web site: http://www.elsevier.nl/inca/publications

Nature Conservancy
The Nature Conservancy
4245 North Fairfax Drive, Suite 100
Arlington, VA 22203-1606
Telephone: (800) 628-6860
Web site: http://nature.org/aboutus/magazine

OnEarth
National Resources Defense Council
40 West 20th Street
New York, NY 10011
Telephone: (212) 727-2700
Web site: http://www.nrdc.org/onearth/02win/default.asp

Organization and Environment
Sage Publications Limited
6 Bonhill Street
London
United Kingdom EC2A 4PU
Telephone: 44 (0) 20 7374 8741
Web site: http://www.sagepub.co.uk

Outside Magazine
400 Market Street
Santa Fe, NM 87501
Telephone: (800) 678-1131
Web site: http://www.outsidemag.com

Ranger Rick
National Wildlife Federation

11100 Wildlife Center Drive
Reston, VA 20190-5362
Telephone: (800) 822-9919
Web site: http://www.nwf.org/rangerrick

Resources
Agricultural Communications, NMSU
Bulletin Office
P.O. Box 30003, MSC 3AI
Las Cruces, NM 88003-8003
Telephone: (505) 646-2701
Web site: http://www.cahe.nmsu.edu/pubs/resourcesmag

Sierra
Sierra Club
85 Second Street, Second Floor
San Francisco, CA 94105-3441
Telephone: (415) 977-5500
Web site: http://www.sierraclub.org/sierra

Society and Natural Resources
UK Head Office
11 New Fetter Lane
London
United Kingdom EC4P 4EE
Telephone: +44 (0) 20 7583 9855
Web site: http://www.tandf.co.uk/journals/tf/08941920.html

Journal, Magazine, and Newspaper Articles

"Alaska Teens Win Fight over Pesticides in Anchorage Schools."
2000. *International Wildlife* 30, no. 1: 8.

Alexander, Charles P. 2001. **"For the Birds."** *Time* (4 June): 66–67.

Austin, Regina, and Michael Schill. 1991. **"Black, Brown, Poor, and Poisoned: Minority Grassroots Environmentalism and the Quest for Eco-Justice."** *Kansas Journal of Law and Public Policy* 1, no. 1: 69–82.

Barnum, Alex. 1998. **"Truck Industry Fights Diesel Plan."** *San Francisco Chronicle* (31 July): A23.

Beck, Melinda, and Michael Reese. 1979. **"Sagebrush Revolt."** *Newsweek* (17 September): 39.

Berke, Richard L. 1999. **"Sierra Club Ads in Political Races Offer a Case of 'Issue Advocacy.'"** *New York Times* (24 October): A-12.

Berlet, Chip, and William K. Burke. 1991. **"Corporate Fronts: Inside the Anti-Environmental Movement."** *Greenpeace* (January–February): 8–12.

Bilman, Jon. 2001. **"Snowmobiles and Yellowstone: Sledheads vs. Greenies."** *Outside Magazine.* http://web.outsidemag.com/environment/wilderness/yellowstone_update.html. Reaccessed 20 August 2002.

Bogo, Jennifer. 2001. **"Green Warriors."** *E Magazine* 12, no. 3 (May–June): 14–18.

Braxton Little, Jane. 1999. **"Crime for Nature."** *American Forests* 105, no. 1 (Spring): 7–10.

Brick, Phil. 1995. **"Determined Opposition: The Wise Use Movement Challenges Environmentalism."** *Environment* 37, no. 8 (October): 38.

Brough, Holly. 1990. **"Minorities Redefine 'Environmentalism.'"** *WorldWatch* (September–October): 5–8.

Bullard, Robert D. 1994. **"The Environmental Justice Movement Comes of Age."** *Amicus Journal* 16, no. 1 (Spring): 32–37.

Callison, Charles. 1980. **"'Sagebrush Rebellion' Is Just Another Name for a Public Lands Heist."** *National Parks and Conservation Magazine* 54 (March): 10.

Cooper, Helene. 1999. **"Some Hazy, Some Erudite and All Angry, WTO Protestors Are Hard to Dismiss."** *Wall Street Journal* (30 November): 1.

Coyle, Marcia. 1992. **"When Movements Coalesce."** *National Law Journal* (21 September). 10 pp.

Dowdle, Barney. 1984. **"Perspectives on the Sagebrush Rebellion."** *Policy Studies Journal* 12, no. 3: 473–482.

"Earth Day 2000: Heroes for the Planet." *Time Magazine Special Edition* (Spring).

Fisher, Julie. 1994. **"Third World NGOs: A Missing Piece to the Population Puzzle."** *Environment* 36, no. 7 (September): 6–11.

Fleming, Donald. 1972. **"Roots of the New Conservation Movement."** *Perspectives on American History* 6: 7–91.

"Grass-Roots Activists: Taking a Stand." 2001. *International Wildlife* 31, no. 5: 3–4.

Hays, Samuel P. 1982. **"From Conservation to Environment."** *Environmental Review* 6 (Fall): 37.

Helvarg, David. 1994. **"Grassroots for Sale."** *Amicus Journal* (Fall): 24–29.

Hornblower, Margot. 1979. **"The Sagebrush Revolution."** *Washington Post* (11 November): B3.

———. 1998. **"Next Stop, Home Depot."** *Time* (19 October): 70.

Ivey, Mark. 1990. **"The Oil Industry Races to Refine Its Image."** *Business Week* (23 April): 98.

Jasanoff, Sheila. 1997. **"NGOs and the Environment: From Knowledge to Action."** *Third World Quarterly* 18, no. 3 (December): 579–594.

Knox, Margaret. 1991. **"Meet the Anti-Greens."** *The Progressive* 55, no. 10 (October): 21–23.

LaBudde, Nathan. 1998. **"Baja's Whales vs. Mitsubishi."** *Earth Island Journal* 13, no. 3 (Summer): 39–42.

Leshy, John D. 1990. **"Unraveling the Sagebrush Rebellion: Law, Politics, and Federal Lands."** *University of California, Davis Law Review* 14: 319–320.

Levy, David L., and Peter Newell. 2000. **"Oceans Apart? Business Responses to Global Environmental Issues in Europe and the United States."** *Environment* 42, no. 9 (November): 8–19.

Markels, Alex. 1999. **"Backfire."** *Mother Jones* (March–April): 60–64, 78–79.

Marston, Ed. 1997. **"The Timber Wars Evolve into a Divisive Attempt at Peace."** *High Country News* (29 September): 1.

Merchant, Carolyn. 1984. **"Women of the Progressive Conservation Movement."** *Environmental Review* 8, no. 1: 57–85.

Moore, Lisa J. 1992. **"When Landowners Clash with the Law."** *U.S. News and World Report* (6 April): 80–81.

Nijhuis, Michelle. 1999. **"ELF Strikes Again."** *High Country News* (1 February): 3.

O'Brien, Anne. 1999. **"Purchasing Power: Why We Still Buy Land."** *Nature Conservancy* (November–December): 12–17.

Paterson, Kent. 2000. **"Mexican Environmentalists behind Bars."** *The Progressive* 64, no. 3 (March): 21.

Pell, Eve. 1990. **"Buying In: How Corporations Keep an Eye on Environmental Groups That Oppose Them."** *Mother Jones* 15 (April–May): 23–26.

Pickerill, Jerry. 2001. **"Environmental Internet Activism in Britain."** *Peace Review* 13, no. 3: 365–370.

Popper, Frank J. 1984. **"The Timely End of the Sagebrush Rebellion."** *Public Interest* 76: 61–73.

Rauber, Paul. 1994. **"Down on the Farm Bureau."** *Sierra* 79 (November–December): 32.

Reiger, George. 1995. **"Good Vibes."** *Field and Stream* (April): 16–17, 22.

Rojas, Aurelio. 1998. **"Rock Climbers Sue over New Yosemite Project."** *San Francisco Chronicle* (28 May): A19.

Russel, Dick. 1989. **"Environmental Racism."** *Amicus Journal* (Spring): 22–32.

Sawhill, John C. 1996. **"The Nature Conservancy."** *Environment* 38, no. 5 (June): 43–44.

Schneider, Paul. 1990. **"When a Whistle Blows in the Forest."** *Audubon* (July): 42–49.

Shimberg, Steven. 2000. **"The Gentle Warrior: John Chafee."** *The Environmental Forum* (January–February): 36–40.

Slater, Daska. 2001. **"Moments of Truth."** *Sierra* (July–August): 48–58.

Steiner, Andy. 2001. **"Vandana Shiva: An Indian Physicist Who Fights for Small Farmers."** *Utne Reader* (November–December): 72.

Tesh, Sylvia, and Bruce Williams. 1996. **"Identity Politics, Disinterested Politics, and Environmental Justice."** *Polity* (Spring): 285–305.

Tuler, Seth, and Thomas Webler. 1999. **"Voice from the Forest: What Participants Expect of a Public Participation Process."** *Society and Natural Resources* 12: 437–453.

Webler, Thomas, Seth Tuler, and Rob Krueger. 2001. **"What Is a Good Public Participation Process? Five Perspectives from the Public."** *Environmental Management* 27, no. 3: 435–450.

Wyss, Jim. 2001. **"Building a Better Banana."** *Amicus Journal* (Winter): 24–25.

Zinsmeister, Karl. 1993. **"The Environmentalist Assault on Agriculture."** *Public Interest* 112 (Summer): 92.

Videotapes

An African Recovery. 28 min. First Run/Icarus Films. 1988.

The catastrophic drought that ravaged parts of Africa in the 1980s served as a catalyst for the citizens of Niger. African initiatives to solve the desertification problem focus on the silted and shallow Niger River.

Alaska: The Last Frontier? 27 min. Annenberg/CPB Collection. 1996.

This program shows the difficulties of balancing the needs of indigenous peoples and the wilderness with economic development and modern life in the state of Alaska.

Anishinaabe Niijii: Friends of the Chippewa. 48 min. A. Gedicks. 1994.

This video documents the opposition, because of possible endangered resources, against a copper mine proposed near the Flambeau River in Ladysmith, Wisconsin.

The Ark Series. 60 min. First Run/Icarus Films. 1993.

Activists protest the actions of London's Regent's Park Zoo as a cost-management team moves in to slash expenses, including ninety people and 1,200 animals—40 percent of the zoo's stock. A rebellious group calls for a takeover of the zoo as less glamorous animals are removed in place of a panda brought from China to breed.

The Ash Barge Odyssey. 58 min. Michael Thomas Productions. 2001.

Using firsthand accounts and supporting material, this documentary tells the story of 14,000 tons of toxic incinerated ash that left Philadelphia in 1986 on a cargo vessel headed for Haiti, where the ash was illegally dumped. It remained there for ten years until efforts by Greenpeace and other environmental groups arranged to have it brought back to the United States.

At the Edge of Conquest: The Journey of Chief Wai-Wai. 28 min. Filmakers Library. 1992.

The Brazilian government proposed reducing the landholdings of the Waipai Indians in Brazil, whose lifestyle is threatened by invading gold miners and the state government's plan to construct a highway directly through their territory. The film shows Chief Wai-Wai's journey to Brazil's capital to navigate through the political system to defend his people and their destiny.

The Banana Verdict. 50 min. Filmakers Library. 2000.

This film focuses on the international banana trade, dominated by major companies such as Chiquita, Dole, and Del Monte. The use of large amounts of pesticides by these companies causes serious illnesses to workers and endangers the environment. At the urging of small-scale banana producers, a European initiative called Agrofair begins to produce organic banana crops as a response to the corporate interests.

Black Water. 28 min. First Run/Icarus Films. 1990.

As pollution from upstream factories and mills flows down to the sea, a maritime community in Bahia, Brazil, is no longer able to sustain the marine life that serves as a food stock and commercial enterprise. Local villagers further deplete stocks as they suffer the human costs of poorly planned industrial development.

Blowing the Whistle: How to Protect Yourself and Win. 35 min. Video Project. 1994.

This training video teaches potential whistle-blowers their rights, demonstrates effective methods of whistle-blowing, and explains what assistance is available to activists. Several whistle-blowers are profiled to show their successes and failures.

Blowpipes and Bulldozers. 60 min. Bullfrog Films. 1989.

This is a film about the Penan, a unique tribe of rain forest nomads residing in Sarawak, Borneo. A Swiss man who lived with the Penan began to see their lifestyle destroyed by logging practices and chronicled the efforts to organize resistance and publicize the Penans' plight.

Borderline Cases. 65 min. Bullfrog Films. 1997.

The factories located along the border between the United States and Mexico produce toxic pollutants that affect the air and water, a process that has an adverse health effect upon the residents of towns on each side. Mexican and U.S. activists are shown documenting the effects of the factories and attempting to gain compliance with environmental regulations

Bound by the Wind. 40 min. Green TV. 1991.

This film investigates the human and environmental impacts of nuclear testing and highlights the international campaign by activists for a comprehensive test ban.

Breaking the Bank. 72 min. Earthfilms. 2000.

In an effort to reclaim media democracy, this film shares the voices and concerns of tens of thousands from all over the world who gathered in Washington, D.C., to protest at meetings of the World Bank and International Monetary Fund. Activist voices are heard uncensored. Representing labor, the environment, civil society, international law, and indigenous nations, this film documents their demonstrations and protests.

The Buffalo War. 57 min. Bullfrog Films. 2001.

The yearly slaughter of America's last wild bison near Yellowstone National Park divides Native Americans, ranchers, government officials, and environmentalists. The film shows the Lakota Sioux, who make a 500-mile spiritual march across Montana to object to the slaughter. Woven into the film are the civil disobedience and activism of an environmental group trying to save the buffalo, as well as the concerns of ranching families caught in the middle.

Can Tropical Rainforests Be Saved? 115 min. Richter Productions. 1999.

This documentary film provides a perspective on the human dimension—indigenous peoples, environmental group leaders, and government officials—involved in a dozen countries in Asia, Africa, and Latin America that have rain forests.

The Cars That Ate Your Life. 30 min. Flying Focus Video Collective. 1992.

Activists in Portland, Oregon, try to wean drivers away from their automobiles and get them to support nonpolluting transportation.

Chemical Kids. 60 min. Filmakers Library. 2002.

This film investigates the alarming effects chemicals have on the human and natural world, focusing on children. Some individuals and communities are fighting chemical plants through demonstrations and lawsuits. Others try to have some effect on the way industrial plants are designed to lower the risk of chemical leakage.

Common Ground: The Battle for Barton Springs. 28 min. Video Project. 1994.

When an exploding population threatens a community's shrinking natural environment, can anything be done? *Common Ground* is an uplifting story that shows how the citizens of Austin, Texas, used the democratic process to protect their endangered natural springs in a positive example for social activists.

Connect: A New Ecological Paradigm. 23 min. Video Project. 1997.

Connect carries one of the most important messages of our time: young people can make a difference in improving the world's environmental and social problems. This film looks at young activist leaders who speak boldly about their power to make a difference for the earth.

The Cost of Cool: Youth, Consumption, and the Environment. 27 min. Video Project. 2001.

American teens from around the nation discuss the burgeoning issue of consumerism and its acute environmental impact. *The Cost of Cool* looks at everyday items, from T-shirts to sneakers, and tracks the effect of their manufacture on the world's resources. Teenagers examine their learned buying patterns, recognizing that much of the stuff they acquire is not needed.

The Cowboy in Mongolia. 51 min. First Run/Icarus Films. 1989.

Dennis Sheehy, a Vietnam veteran, moves his family to the steppes of Inner Mongolia to help herders whose animals are turning the once lush grassland into desert. He works to convince the traditional Mongols that the future of their region depends upon the sharing of knowledge and trust between representatives of differing cultures.

Deadly Deception. 29 min. Infact. 1991.

This exposé of the human and environmental effects of General Electric Company's nuclear weapons facilities shows plant workers poisoned by radiation and asbestos. Citizens in neighboring homes have also experienced cases of cancer and birth defects. The film shows the activists who are working to inform the public and stop the company's activities.

Decade of Destruction: The Crusade to Save the Amazon Rain Forest. 5-part series. 272 min. Bullfrog Films. 1990.

This documentary film series depicts the systematic destruction of the Amazon rain forest from 1980 to 1990. Each episode follows the life stories of people caught up in the frontier's web of need and greed, stories of personal tragedy and great courage. The programs relate the individuals' struggles to the wider developments going on around them. Together they illustrate the principal issues of Amazonia during the 1980s—its decade of greatest destruction.

Democratic Allsorts. 53 min. Bullfrog Films. 1993.

This is a portrait of Frances Moore Lappe, author of the well-known book, *Diet for a Small Planet,* and how she continues her efforts to prevent hunger and starvation by focusing on democratic decision making on an individual and global level.

Devils Tower. 30 min. Bullfrog Films. 2001.

Rock climbers have used the Devils Tower for decades, even though the surrounding environment is used by the Lakota Indians for the performance of sacred dances and vision quests. The Indians' efforts are shown in this film as they explain how rock climbers are disturbing the site and their native cultural practices.

EarthFirst: The Struggle for the Australian Rainforest. 58 min. Video Project. 1989.

This film highlights the dramatic fight to preserve the last stand of majestic rain forests in Australia. It includes commentary from leading scientists on the importance of rain forests and environmental protection.

Edward Abbey: A Voice in the Wilderness. 58 min. Canyon Productions. 1993.

This documentary highlights one of the American West's most passionate advocates. Abbey, who died in 1989, wrote the activist novel, *The Monkey Wrench Gang,* which inspired the group Earth First! Through archival footage and interviews with Abbey's friends and family, the film is both a tribute and a biography.

El Dorado. 56 min. Video Project. 1996.

This is an emotional story of a community in conflict over its natural resources. The film takes an unusually unbiased look at four local residents—two timber workers and two environmentalists—as they try to balance the health of the El Dorado National Forest in the Sierra Nevada foothills with the jobs of the workers who depend on it.

Environment. 60 min. Annenberg/CPB Collection. 1994.

This film looks at the international dimension of environmental problems, focusing on transnational pollution, international property rights, and perceived differences between trade and environmental protection. The United States–Mexico agreement on dolphin-safe tuna is explored, as is the transnational implication of pollution for citizens who live along the borders of the Rhine River.

Fat of the Land. 56 min. Video Project. 1996.

Who would have taken a bet that anyone could drive a Chevrolet van across the United States fueled by leftover vegetable oil? Five enterprising young women did just that, traveling 2,300 miles using vegetable oil as an alternative fuel. The activists proved there is a sustainable and renewable alternative to petroleum.

Field of Genes. 44 min. Bullfrog Films. 1998.

Technology has quietly slipped into the food chain, shifting genes from one life-form to another. Multinational chemical companies have created genetically altered potatoes, corn, soybeans, and canola that can be toxic to pets, are herbicide tolerant, and are dependent upon chemical inputs. Environmentalists question the claims of the companies and the impact on the hungry, the farmer, and consumers.

Fire in the Eyes. 32 min. Headwaters Action Video Collective. 1999.

Faced with mounting demonstrations against logging in California's Headwaters redwood forest, Humboldt County law enforcement officers decide to use pepper spray directly in the eyes of nonviolent protesters. This controversial tactic, called "pain compliance," is the subject of this film about environmental activism and police.

Fire in Their Hearts. 46 min. Films for the Humanities and Sciences. 1990.

In profiles of citizen activists, the film looks at citizens such as Lois Gibbs, a homemaker who protested against environmental health hazards in the Love Canal area of upstate New York; a naturalist trying to preserve wildlife and open spaces in California's Marin County; and a retired couple whose religious beliefs compelled them to become nuclear disarmament activists.

Fisheries: Beyond the Crisis. 46 min. Bullfrog Films. 1998.

As fisheries around the world reach crisis levels from overharvesting, this film examines the effects on two communities, one in Canada's Bay of Fundy and one in South India. Residents organized, demanded control of the fishery industry for their communities, and dismissed a quota system favorable to corporations.

Fishing in the Sea of Greed. 45 min. First Run/Icarus Films. 1998.

This film documents the response of one fishing community in India to the industries that have begun to dominate their livelihood and decimate their environment. Under the leadership of the National Fishworkers Forum and the World Forum of Fishworkers and Fish Harvesters, the workers fight for their jobs and their communities.

Flooding Job's Garden. 59 min. First Run/Icarus Films. 1991.

The Cree people of James Bay are documented in this story of what happens to land claims and self-government years after the development of the region's huge hydroelectric project. The Crees are mounting an international campaign to protect the environment and ensure responsible development.

Food or Famine? 49 min. Filmakers Library. 1997.

Increasing global population will lead to more mouths to feed. Agriculture production, which has led to surpluses on world markets, has also led to environmental degradation, such as erosion, salinization, and chemical pollution. This film highlights agriculture projects around the world as farmers work to incorporate methods based on sound ecological principles.

Fooling with Nature. 60 min. PBS Video. 1998.

The program examines new evidence in the controversy over the danger of man-made chemicals to human health and the environment, thirty-five years after Rachel Carson first raised concerns of an impending ecological crisis. The film takes viewers inside the world of the scientists, politicians, activists, and business officials embroiled in this high-stakes debate.

For Earth's Sake. 58 min. Bullfrog Films. 1997.

This is a portrait of the late David Brower, an environmentalist instrumental in establishing some of America's most spectacular national parks. He also headed the Sierra Club, Friends of the Earth, and Earth Island Institute, three of the country's most important environmental organizations. Brower was twice nominated for the Nobel Peace Prize for his efforts.

The Forest for the Trees. 58 min. Green TV. 1990.

This film takes a sobering look at the battle for the redwoods of California. Loggers, timber workers, politicians, environmentalists, and local residents all speak passionately of their concerns and interests.

Forest Wars: The Fight for Coolangubra. 60 min. Media Associates. 1995.

This film is a classic resource story that graphically depicts what is at stake on all sides. It looks at how citizens with different viewpoints and interests are affected by, and try to influence, government policy.

Fury for the Sound: The Women of Clayoquot. 86 min. TellTale Productions. 1997.

The important role of women activists in establishing grassroots social movements is seen in this film about the protests against clear-cut logging on Canada's West Coast. It depicts women of all ages fighting to protect their region, one of the largest remaining tracts of untouched Canadian rain forest.

Going Green. 22 min. Bullfrog Films. 1992.

As an example of individual activism, one family reduces their waste to only one and a half garbage bags per year. The film identifies ways in which items can be recycled, alternatives to household chemicals, and the benefits of composting and buying in bulk.

The Golf War. 39 min. Bullfrog Films. 2000.

A Philippine government plan to transform ancestral farmland into a tourist resort sparks a dramatic conflict when villagers actively resist the development. It is a provocative portrait of one community's fight for survival against forces of economic development.

Good Wood. 45 min. Bullfrog Films. 1999.

Communities that depend upon the forest for their livelihood are seen in this film, which shows how citizens are discovering vital links that keep people employed while simultaneously preserving the world's forests.

Handle with Care. 26 min. Bullfrog Films. 1992.

This film highlights two activists who are fighting industrial development that damages the environment and endangers human health. Jorunn Eikjok, a Norwegian woman and a member of the Sami reindeer herders in the Arctic Circle, and Lois Gibbs, an American woman who has taken on some of the most powerful companies in the United States, are depicted.

Healing the Earth. 28 min. National Geographic Society. 1995.

How can we help heal the earth through cooperation between communities? That question is addressed as residents deal with Superfund projects and the use of bioremediation.

Heart of the People. 58 min. University of California Extension Center for Media. 1997.

The film documents the Huu-ay-aht native people of Vancouver Island, Canada, who are taking a nonconfrontational approach to reclaiming a river devastated by logging and restoring it and its resources for future generations.

Henry's Way. 53 min. Bullfrog Films. 1998.

This is a portrait of Henry Spira, the most effective animal rights activist of the past twenty years. It documents Henry's foresight in proposing positive outcomes for his opponents while at the same time advancing his own cause.

Heroes of the Earth. 45 min. Video Project. 1993.

There is a new kind of hero: people who are fighting to protect the environment, often at great personal risk. This film profiles seven recent Goldman award winners who are exemplary of individual action in the critical issues of environmentalism.

Hopi Land. 29 min. Bullfrog Films. 2001.

The Hopi tribe is an indigenous community near the Four Corners area of the United States who are struggling to protect their land from companies such as Peabody Coal. The firm is pumping millions of gallons of water daily from a local aquifer in order to transport its coal to a processing plant in Nevada.

Human Faces behind the Rain Forest. 30 min. First Run/Icarus Films. 2001.

This film documents the events surrounding the harvest of the opium poppy crop in the Colombian rain forest, showing how indigenous communities are forced by social and economic conditions into the poppy trade. It profiles individuals who, despite death threats, are trying to eradicate the poppy flower and shows how various peasant groups are seeking alternatives.

In Our Own Backyard. 59 min. Bullfrog Films. 1983.

Documenting the nation's first encounter with toxic waste, the film highlights the moving account of the people involved, from the community members to the government. It shows how local citizens, led by Lois Gibbs, developed strategies including public protest and skillful use of the media to influence the government's response.

In the Light of Reverence. 75 min. Bullfrog Films. 2001.

Across the United States, Native Americans are struggling to protect their sacred places. Religious freedom, so valued in the United States, is not guaranteed to those who practice land-based religions. The film profiles three indigenous communities in their struggles to protect their sacred sites from rock climbers, tourists, strip mining, development, and New Age religious practitioners.

Itty, Bitty, City: Bradshaw on Walking. 30 min. Flying Focus Video Collective. 1995.

Chris Bradshaw, pedestrian authority and founder of Ottawalk, Canada's main society for sidewalk users, gives his views for creating a more walkable city.

Judi Bari vs. the United States of America. 30 min. Flying Focus Video Collective. 1994.

Famed activist Judi Bari, a leader in the Earth First! movement, is profiled in this documentary about her arrest in what she believes was a government cover-up of a car bombing that severely injured her and another activist.

Keepers of the Coast. 31 min. Bullfrog Films. 1996.

With a rock music background, this films shows the work of local surfers and the Surfrider Foundation to educate the public and protest the pollution of coastlines.

Keepers of the Water. 39 min. A. Gedicks. 1996.

This documentary film shows the struggle to keep the headwaters of the Wolf River in Wisconsin from being polluted by a proposed sulfide mine. The area is also home to the Sokaogon Mole Lake Chippewa tribe.

Killing Coyote. 83 min. High Plains Films. 2000.

It is estimated that 400,000 coyotes are trapped, shot, and poisoned in the United States every year; many of them are killed using taxpayer dollars. Others are killed for fun, cash, and prizes, but the coyotes continue to survive. Ranchers are trying to protect their livestock, activists are trying to protect the coyote, and politicians try to listen to each group's concerns.

Laid to Waste: A Chester Neighborhood Fights for Its Future. 52 min. University of California Extension Center for Media. 1996.

This film documents the residents of Chester, Pennsylvania, who are fighting against the location of another waste treatment plant in their city. Through community action, residents demand that they be heard.

The Last Rivermen. 31 min. Riverkeeper. 1997.

The anglers and their families who live along New York's Hudson River join together in a unique grassroots effort to stop the chemical pollution that has ruined their livelihood in one of the first examples of environmental activism in the United States.

The Last Stand: Ancient Redwoods and the Bottom Line. 57 min. University of California Extension Center for Media. 2002.

The destruction of the ancient redwoods of northern California highlights testimony from economists, scientists, forest activists, and the Humboldt County residents who have seen their communities and backyards torn apart.

The Last Stand: The Struggle for the Ballona Wetlands. 57 min. Video Project. 1999.

This documentary covers the struggle over the Ballona Wetlands ecosystem, where eighty-seven local citizens' groups fight Playa Vista, the largest development in the United States. There are interviews with people on various sides of the issue who relate their concerns.

Living with the Spill. 52 min. First Run/Icarus Films. 1990.

The 1989 *Exxon Valdez* oil spill in Alaska cost the corporation more than $1.5 billion in an effort to restore the area's ecosystem.

But little was done to heal the human misfortune that resulted, including soaring property prices, the closing of the salmon cannery, and an infusion of enormous amounts of new money that tore the fabric of the once peaceful community apart.

Logs, Lies, and Videotape. 12 min. Green Fire Productions. 1996.

Filmed in the American Northwest, this video looks at the impact of a logging measure passed by Congress that directs the U.S. Forest Service and the Bureau of Land Management to accelerate salvage logging and its impact on the communities nearby.

Lost Forever: Everett Ruess. 60 min. Southern Utah Wilderness Alliance. 2001.

As an ambitious young artist, Everett Ruess left Los Angeles in 1934 to hitchhike to Arizona. He spent his time in the canyons of the Southwest, abandoning his artistic career to become closer to nature. He vanished in the wilds of southern Utah and became a mysterious icon in American environmental history, symbolic of those seeking the deep peace of the wilderness.

Luna: The Stafford Giant Tree Sit. 20 min. Earthfilms. 2001.

The giant redwood tree, also called Luna, was discovered by Earth First! activists during the fall 1997 Headwaters campaign. The tree was blue-marked by the lumber company for cutting. In order to save it from the chain saw, a nonstop tree occupation began. One extraordinary woman named Julia Butterfly remained in the tree to protest logging in ancient forests.

Mining Seven-Up Pete. 31 min. High Plains Films. 1995.

This film investigates the struggles of Montanans to save the famed Blackfoot River in western Montana from what would be one of the largest cyanide-heap leach gold mines in North America. It documents the effects of mining, from cyanide-laced groundwater to acid mine drainage.

The Monk, the Trees, and the Concrete Jungle. 26 min. Bullfrog Films. 1993.

Focusing on Southeast Asia, this program looks at what is being done in Thailand to stop deforestation and the trade in endangered species. It also looks at Japan, a major consumer of tropical hardwood, to see what actions are being taken to change people's attitudes and influence government policy, with a focus on the Japanese Tropical Forest Action Network.

A Narmada Diary. 60 min. First Run/Icarus Films. 1995.

The Save Narmada Movement, or Armada Bachao Andolan, has emerged as one of the most dynamic environmental struggles in India, with nonviolent protests by residents who would prefer to drown rather than to leave their homes and land.

The Naturalist. 32 min. High Plains Films. 2001.

Kent Bonar, who has been called "the John Muir of the Ozarks," is one of America's great naturalists. Living without modern amenities in the style of Henry Thoreau, Aldo Leopold, and John Muir, Bonar has spent his life observing and recording the natural history of the region.

The Next Industrial Revolution. 55 min. Earthome Productions. 2000.

In a documentary format, Bill McDonough explains his vision in which humanity takes nature itself as a guide and his efforts to create a movement that reinvents technical enterprises.

North Portland Concerns. 30 min. Flying Focus Video Collective. 1992.

Residents in the St. Johns neighborhood in Portland, Oregon, organize to battle a proposed waste incinerator that is slated to be built in their midst.

Nuclear Dynamite. 52 min. Bullfrog Films. 2001.

This film explores the Soviet-American race to develop nuclear weapons and the testing of more than 150 nuclear weapons between 1958 and 1988. The testing finally halted owing to the emergence of environmental movements and protests in both countries.

Peacock's War. 60 min. Bullfrog Films. 1989.

Doug Peacock, the man who served as the model for Hayduke in Edward Abbey's *The Monkey Wrench Gang,* is profiled in the story of one man's determination to resist the destructive forces that threaten our world and the survival of the wild creatures that live in it.

Politics, People, and Pollution. 58 min. Public Affairs Television. 1992.

Through interviews with activists, including Lois Gibbs of New York's Love Canal, the video explains the concerns of citizens, especially those who live in Louisiana's "chemical alley."

A Question of Power. 58 min. Video Projects. 1986.

This is an important in-depth chronicle of one of the most important struggles over nuclear power—the lengthy effort to prevent the start-up of the Diablo Canyon nuclear facility in California.

Rachel Carson's Silent Spring. 57 min. PBS Video. 1993.

Called one of the most important books of the twentieth century, Rachel Carson's famous book *Silent Spring* identified the legacy of the poisoning of the environment. This one scientist's courage and determination changed the way we think about the world and served as the catalyst for the environmental movement.

Radioactive Reservations. 52 min. Filmakers Library. 1995.

Tribal leader Eagleye Johnny travels to four Indian reservations in Oregon, Utah, New Mexico, and Nevada to discuss the residents' negotiations with the U.S. government to place Monitored Storage Retrieval sites for radioactive waste from nuclear plants on their land.

Rage over Trees. 58 min. PBS Video. 1989.

This video provides an introduction to the head-to-head struggle over the Pacific Northwest's last ancient forests and discusses the activities of environmental organizations, such as Earth First!, as they demonstrate against logging in Oregon.

Redwood Summer. 30 min. Bullfrog Films. 1991.

In vivid footage, the protests and civil disobedience against timber companies harvesting ancient redwoods in northern California are documented. The activities of the group Earth First! are highlighted to show the conflicts among community members, loggers, and activists.

Regopstaan's Dream. 52 min. First Run/Icarus Films. 2000.

The last surviving South African Bushmen tell their story in their fight to reclaim ancestral land in the Kalahari Gemsbok National Park. This struggle encompasses the sensitive issues of wildlife conservation and the rights of indigenous people to live in protected areas.

River of Broken Promises. 24 min. Environmental Health Coalition. 1996.

The New River, which runs through the mid-Atlantic states region, considered the dirtiest river in the United States, has had little governmental attention over the last fifty years. Narrated by Martin Sheen, this film examines the causes of contamination and the community efforts to clean up the toxic waterway, a cleanup that has had a low priority, as an environmental justice issue.

Road Use Restricted. 18 min. Whispered Media. 1997.

Earth First! activists block a logging road in Idaho to attempt to save the largest wilderness area left in the lower forty-eight states. The protesters encounter federal agents and a raid on a makeshift road blockade.

Showdown in Seattle: Five Days That Shook the WTO. 150 min. Earthfilms. 1999.

This video features an on-the-ground, noncorporate perspective of police action and popular resistance during the World Trade Organization (WTO) demonstrations in Seattle. The film was shot and edited on location by a collaboration of video producers and was originally shown each day of the WTO protests.

Since the Company Came. 52 min. First Run/Icarus Films. 2000.

When village leaders invite a Malaysian company to log their tribal land, the Haporai people of Rendova island in the South Pacific find themselves confronted with a choice between earning money and preserving the forests and traditions that sustain their families.

The community's leaders are desperate to stop the logging, as the village's women argue with the men over traditional social values.

SOS Sarajevo: War Hurts Animals Too. 30 min. Flying Focus Video Collective. 1999.

The war in Sarajevo takes its toll on humans as well as animals in this documentary footage that focuses on the efforts of groups and individuals to save them.

Southbound. 48 min. High Plains Films. 1995.

As the wood products industry races to feed global demand for paper and lumber, citizens are fighting to protect remnants of the recovering hardwood forests of the South. *Southbound* goes to the heart of the struggle through interviews representing many sides of the issue.

Stepan Chemical: The Poisoning of a Mexican Community. 18 min. The Coalition. 1992.

The film shows how the dumping of hazardous wastes by a Mexican subsidiary of Stepan Company, based near Chicago, has contaminated communities in Matamoros, Mexico. Community leaders, with the help of the Coalition for Justice in the Maquiladoras, demand an end to contamination and a full accounting of the company's activities.

Sustainable Futures. 38 min. Video Project. 1999.

Sustainable Futures profiles seven locations in three countries where there are demonstrations of local community efforts toward sustainability. French farmers at Ruisseau St. Esperit, Canada, are featured along with El Paso, Texas, activists worried about pesticides and Chicago's Friends of the River seeking to clean up their waterways.

Taking the Waters. 26 min. Bullfrog Films. 1993.

This film looks at two critical wetland sites—one in the Danube Basin, and the other in Tunisia—and the people who are trying to turn each area into a national park. One is a homemaker turned environmental warrior; the other a government botanist who challenges his employers.

Thinking Green. 35 min. The Greens/GPUSA National Clearing-house. 1994.

The Green Party of the United States explains the links between the feminist movement and the green movement, as well as other issues related to the environment.

Thinking like a Watershed. 27 min. Video Project. 1998.

A small, coastal community in northern California has been restoring its watershed for the past twenty years. Originally determined to save the nearly extinct native salmon, they realize that salmon do not just live in streams, they live in watersheds.

This Land Is Your Land. 15 min. High Plains Films. 1999.

This video investigates a long-standing management practice on national forest lands of selling timber from federal lands to the highest bidder. It features former U.S. Forest Service employees, biologists, conservationists, and two members of Congress who question the policy.

Timber GAP. 18 min. Headwaters Action Video Collective. 2000.

Judi Bari, an organizer for the group Earth First!, explains how timberland can be converted once the forest has been clear-cut, showing how vineyards and gentrification are moving north into the redwood region. The film shows protests and civil disobedience by a variety of activists, including footage of World Trade Organization demonstrations in Seattle and against the family of David Fisher, founder of the clothing company, the Gap.

Toxic Racism. 60 min. Films for the Humanities. 1994.

This exploration of the grassroots war against those responsible for the high levels of toxic waste dumping and industrial pollution deals with poor minority neighborhoods.

Toxic Waters. 58 min. Michael Thomas Productions. 2000.

In the early 1950s, trucks loaded with trash began dumping their cargo in a swampy site now called the Clearview Landfill in Pennsylvania. Years later, residents band together when they discover much of the waste is toxic and begin "the good fight" against political influence.

Tree Sit: The Art of Resistance. 120 min. Earthfilms. 2001.

Demonstrators who are protesting the logging of the ancient red-wood forests of northern California, owned by the MAXXAM Corporation, look to new strategies and tactics. In an example of direct action, they decide to build platforms in the trees in order to save them from being logged.

Trinkets and Beads. 53 min. First Run/Icarus Films. 1996.

This video documents the lives of the Huaorani, a small tribe of Ecuadorean Indians, who after twenty years of pressure from foreign oil companies agreed to allow oil drilling on their land. It focuses on the introduction of massive environmental pollution and cultural change and the tribe's subsequent efforts to regain control of their lives and land.

Txai Macedo. 50 min. First Run/Icarus Films. 1992.

Antonio Macedo leads an alliance of Indian and white rubber tappers in the Brazilian rain forest against drug lords, rubber companies, landowners, and a corrupt legal system. The film shows Macedo at work, furthering his people's cause, even when attempts are made to kill him.

Urban Environment. 60 min. EcoVideo. 1999.

This is a discussion of the work of environmental activists in an urban setting, including Bernadette Kozart, who is one of the leaders of the Greening of Harlem, and Andy Lipkis of Los Angeles's TreePeople.

Voice of the Amazon. 57 min. Cinema Guild. 1989.

Through interviews and documentary footage, this film examines the rubber tapper movement in Brazil. It profiles Chico Mendes, assassinated in 1988 for his role in trying to protect the rain forest from exploitation.

Voices from the Frontlines. 38 min. Cinema Guild. 1997.

The film profiles the Los Angeles–based Community Strategy Center, a grassroots nonprofit organization that is deeply involved in the city's environmental issues.

Voluntary Simplicity. 23 min. Filmakers Library. 1999.

This new environmental movement seeks to question the current consumption-driven definition of values and identity. The film tracks the lives of two individuals who have voluntarily begun to simplify their lifestyles.

The Waterkeepers. 48 min. First Run Features. 1999.

Robert F. Kennedy Jr. hosts this documentary coverage of dozens of river, bay, and sound keepers from Alaska to North Carolina. The resolve of the environmentalists is seen in their efforts to protect open waters.

Whose Home on the Range? 55 min. Bullfrog Films. 1999.

This film investigates the environmental conflict in Catron County, New Mexico, that pits ranchers, loggers, environmental activists, and the U.S. Forest Service against each other over land use. Health issues are used as a catalyst to bring the parties together.

The Wilderness Idea. 58 min. Direct Cinema Limited. 1990.

The American conservation movement started with John Muir and Gifford Pinchot and the idea of protecting and using wild areas. The two men battle over the use of California's Hetch Hetchy Valley in Yosemite National Park—a turning point in American environmentalism.

Wildland. 35 min. High Plains Films. 2000.

Natural landscapes are disappearing faster than ever before in the United States. Urban sprawl, logging, mining, and road building threaten to eliminate the last wild areas. Citizens unite in their communities to preserve the last wild places, from tiny wetlands to vast desert expanses.

Wind River. 34 min. High Plains Films. 1999.

Documentary filmmaker Drury Gunn Carr describes the battle of the Shoshone tribe in Wyoming to change the state's water laws. In 1991, the tribe sued to be allowed to send water downstream to fish runs, eventually losing its case to sugar beet farmers who diverted water for their fields.

You Bet They Die. 60 min. Flying Focus Video Collective. 1999.

Animal rights activists document the 20,000 greyhounds who are put to death each year because they can no longer win races. The film identifies what individuals can do to stop the abuse of the racing dogs.

Video Distributors

Note that some videotapes are available through more than one distributor, and prices will vary depending upon the source. Many small distributors will make their films available at no or minimal cost to schools, colleges, and nonprofit organizations upon request.

A. Gedicks
210 Avon Street, #4
La Crosse, WI 54603
(608) 784-4399

Annenberg/CPB Collection
P.O. Box 2345
South Burlington, VT 05407-2345
(800) 532-7637

Bullfrog Films
P.O. Box 149
Oley, PA 19547
(800) 543-3764

Canyon Productions
5905 Lenox Road
Bethesda, MD 20817
(301) 229-1836

Cinema Guild
130 Madison Avenue, Second Floor
New York, NY 10016-7038
(212) 685-6242

The Coalition
Coalition for Justice in the Maquiladoras
530 Bandera Road
San Antonio, TX 78228
(210) 732-8957

David L. Brown Productions
274 Santa Clara Street
Brisbane, CA 94005
(415) 255-7469

Direct Cinema Limited
P.O. Box 10003
Santa Monica, CA 90410-1003
(310) 636-8200

Earthome Productions
P.O. Box 212
Stevenson, MD 21153
(410) 902-3400

EcoVideo
988 Cranberry Drive
Cupertino, CA 95014
(408) 865-0888

Environmental Health Coalition
1717 Kettner Boulevard, Suite 100
San Diego, CA 92101
(619) 235-0281

Filmakers Library
124 East 40th Street
New York, NY 10016
(212) 808-4980

Films for the Humanities and Sciences
P.O. Box 2053
Princeton, NJ 08543-2053
(800) 257-5126

First Run Features
153 Waverly Place
New York, NY 10014
(800) 229-8575

First Run/Icarus Films
32 Court Street, Twenty-first Floor
Brooklyn, NY 11201
(718) 488-8900

Flying Focus Video Collective
3439 NE Sandy Boulevard, PMB#248
Portland, OR 97232
(503) 239-7456

Green Fire Productions
33 SE 62nd Avenue
Portland, OR 97215
(503) 736-1295

Green TV, Inc.
1125 Hayes Street
San Francisco, CA 94117-1623
(415) 255-4797

The Greens/GPUSA National Clearinghouse
P.O. Box 1134
Lawrence, MA 01842
(978) 682-4353

Headwaters Action Video Collective/Earthfilms
P.O. Box 2198
Redway, CA 95560
(415) 820-1635

High Plains Films
P.O. Box 8796
Missoula, MT 59807
(406) 543-6726

Infact
46 Plympton Street

Boston, MA 02118
(617) 695-2525

Media Associates
Canberra Business Centre
Bradfield Street
Downer
Australia ACT 2602
06 242 1998

Michael Thomas Productions
P.O. Box 26
Montgomeryville, PA 18936
(610) 294-3036

National Geographic Society
P.O. Box 957
Hanover, PA 17331
(800) 437-5521

PBS Video
P.O. Box 751089
Charlotte, NC 28275
877-PBS-SHOP

Richter Productions, Inc.
330 West 42nd Street
New York, NY 10036
(212) 947-1395

Riverkeeper
25 Wing and Wing
Garrison, NY 10524-0130
(845) 424-4149

Southern Utah Wilderness Alliance
1471 South 1100 East
Salt Lake City, UT 84105
(801) 531-0064

TellTale Productions
1956 East 3rd Avenue

Vancouver, BC
Canada V5N 1H5
(877) 239-7754

University of California Extension Center for Media
2000 Center Street, Suite 400
Berkeley, CA 94704
(510) 642-4124

The Video Project
P.O. Box 77188
San Francisco, CA 94107
(800) 475-2638

Whispered Media
P.O. Box 40130
San Francisco, CA 94140
(415) 789-8484

Electronic Databases

Note that many of the databases listed below are available on a subscription basis only. Most public and university libraries have these databases as part of their collections, however, and they can easily be accessed on-line. Many have tutorials to help users find the information they are seeking. The databases are constantly being updated to include more information and are sometimes consolidated with other databases.

Academic Search Elite

Provides abstracts and indexing for over 3,100 scholarly journals and author-supplied abstracts for 400 of those, covering the social sciences, humanities, education, and more. Also offers full text for over 1,000 journals, including 380 journals dating back to 1990.

Academic Universe (Lexis-Nexis, Congressional Universe)

Academic Universe provides access to a wide range of news, business, legal, and reference information from a wide variety of sources.

AccessUN

An index to United Nations documents and publications.

ACS (American Chemical Society Publications)

Full-text searchable database of all ACS journals.

AGRICOLA

Covers every major agricultural subject, including environmental pollution, pesticides, and water resources. The database includes citations for journal articles, book chapters, monographs, conference proceedings, serials, technical reports, and other materials from more than 5,000 journals.

American National Biography (ANB)

ANB offers portraits of more than 17,400 men and women from all eras and walks of life whose lives have shaped the nation. It is updated quarterly with hundreds of new entries each year. It also includes more than 80,000 hyperlinked cross-references.

Article1st

Contains bibliographic citations and tables of content for approximately 12,600 journals in science, technology, medicine, social science, business, the humanities, and popular culture.

Associations Unlimited

This database includes nonprofit membership organization information, with over 144,000 listings for groups around the world.

Biography Index

Cites more than 2,700 periodicals and 1,800 books, including individual and collective biographies.

Biology Digest

This is a compilation of abstracts and indexes of domestic and international literature in the life sciences. It is primarily intended for an audience at the high school or undergraduate college level. Each of over 20,000 records contains an original abstract averaging 325 words.

Encyclopaedia Britannica Online

Provides an electronic encyclopedia and dictionary, with links to Internet resources and an Internet guide.

Environmental Knowledgebase Online

Published since 1972, this database contains nearly 750,000 indexed citations to articles from scientific, technical, and popular journals, spanning the whole range of environmental topics.

Eventline

A multidisciplinary, multinational database containing information on past and coming events.

Expanded Academic Index (Infotrac)

This source indexes periodicals in all subjects in a wide variety of scholarly and popular publications, some of which are available in full-text format on-line.

FactSearch

This is a guide to statistical statements on current social, economic, political, environmental, and health issues derived from newspapers, periodicals, newsletters, and documents.

Global Newsbank

Provides full-text articles from 1996 on from more than 1,500 sources covering varying perspectives on international issues and events. Includes news stories from radio and television broadcasts, wire services, newspapers, magazines, and government statements, in English, and updated daily.

Historical Abstracts

A reference guide to world history from 1450 on, excluding the United States and Canada; includes indexing for journal articles.

Index to Legal Periodicals and Books

Provides an index of over 890 journals, yearbooks, law reviews, and bar association publications, including court decisions and legislation.

NetFirst

Contains bibliographic citations with summary descriptions and subject headings, describing Internet-accessible resources on a wide range of topics.

NewsBank NewsFile Collection

A full-text news resources database consisting of articles from 1992 on selected from over 500 newspapers and ten news wires. The database provides coverage of regional, national, and international issues along with maps, illustrations, charts, and tables.

PAIS International

This bibliographic index deals with public policy, social policy, and the social sciences. Included are journal articles, books, government documents, statistical compilations, committee reports, and other printed material from all over the world.

Political Science Abstracts

Articles published since 1976 are included in this database, indexed from professional journals, major news magazines, and books.

Statistical Universe

Statistical data produced by the U.S. government; important international intergovernmental organizations; professional, business, and trade organizations; publishers; research organizations; state government agencies; and universities are included.

TreeCD

Citations and abstracts for journal articles, books, and reports on forestry, agroforestry, forest management, tree biology, pest control, and land use.

UnCover

Indexes over 26,000 journals from the 1980s to date. All types of periodicals are included, and virtually every subject is represented.

World Book Encyclopedia

This is an on-line version of the *World Book,* with more than 17,500 full-text articles and more than 6,000 cross-references.

Other Nonprint Resources

Daily Grist
http://www.gristmagazine.com

This site provides daily environmental news via e-mail and includes a humorous perspective on the issues.

Earth Times
http://www.earthtimes.org

There are three separate editions of this publication: a daily Internet edition, a biweekly print edition, and one available by subscription and at newsstands. Published by the nonprofit Media Foundation, the newspaper serves as a "forum for voices from the field."

EE-Link
http://www.eelink.net

The North American Association for Environmental Education sponsors this site, which includes information about organization conferences, classroom resources for K–12 and higher education institutions, grant resources, and links to other groups.

Electronic Green Journal
http://egj.lib.uidaho.edu/

This on-line journal, compiled and distributed through the University of Idaho Library, includes articles on a wide spectrum of issues plus a large number of reviews of environmental books.

Environment News Service
http://www.ens.lycos.com

Developed by the Internet search engine Lycos, and updated daily, this site provides summaries and links to the full article on environmental issues.

Environmental News Network (ENN)
http://www.enn.com

Using the Internet as a communications tool, the ENN, founded in 1993, provides environmental news, live chats, feature stories, and forums for debate on all sides.

Environmental Protection Agency (EPA)
http://www.epa.gov/enviro

This site allows retrieval of EPA information and reports, maps, technical information, and other data from eight EPA databases in what the agency calls its Envirofacts Warehouse.

http://www.epa.gov/OCEPterms

This site defines in nontechnical language the more commonly used environmental terms. It includes acronyms and words used in government reports, by the media, and in public documents.

Green Wave
http://www.greenwaveradio.com

This radio program allows listeners to comment or advertise on its national environmental issues forum.

Headwaters News
http://www.headwatersnews.org

This site provides both a daily and weekly news analysis of environmental issues and activism, with links to news stories in other publications. Its primary coverage is the Rocky Mountain West and the Canadian Rockies.

National Environmental Data Index
http://www.nedi.gov

Federal agency information about environment issues is directly accessible, including reports and statistics from the Department of Agriculture to the National Aeronautics and Space Administration.

National Library for the Environment
http://www.cnie.org/NLE

This is a listing of over 3,000 links, all annotated, to reference resources, databases, and organizations associated with the environment.

Planet Ark
http://www.planetark.org

Updated daily with environmental news from the Reuters news service, this site includes celebrity interviews on environmental issues and a radio broadcast.

Index

About the Author

Jacqueline Vaughn Switzer is an associate professor in the Department of Political Science at Northern Arizona University, where she specializes in public policy and administration. Her environmental background includes work in both the public and private sector, and she has published extensively on a wide range of environmental issues.